WEATHER
OF THE SAN FRANCISCO
BAY REGION

CALIFORNIA NATURAL HISTORY GUIDES

Phyllis Faber and Bruce Pavlik, General Editors

RECENT TITLES IN THE SERIES

GUIDES 63

THE
SAN FRANCISCO
BAY REGION

HAROLD GILLIAM

SECOND EDITION

UNIVERSITY OF CALIFORNIA PRESS

Berkeley   Los

*The publisher gratefully acknowledges the
generous contribution to this book provided by the
Moore Family Foundation
Richard & Rhoda Goldman Fund
and the
General Endowment Fund of the University of California Press Associates*

University of California Press
Berkeley and Los Angeles, California

University of California Press, Ltd.
London, England

© 1962, 2002 by the Regents of the University of California

Photo credits: Frontis—courtesy of the San Francisco Convention and Visitors Bureau; Figure 4—Arthur C. Smith; Plates 1, 3, 4, 25—courtesy of the San Francisco Convention and Visitors Bureau; Plate 2—Sandor Balatoni, courtesy of the San Francisco Convention and Visitors Bureau; Plates 5, 8, 9, 10, 11, 12, 13, 14, 15—Harold Gilliam; Plate 6—courtesy of the Smith Novelty Co.; Plate 7—courtesy of the Golden Gate Bridge Highway and Transportation District; Plates 16, 17, 18, 19, 21, 22—Arthur C. Smith; Plate 20—Bill Sciallo; Plate 23—Regis Lefebure, courtesy of the San Francisco Convention and Visitors Bureau; Plate 24—Carl Wilmington, courtesy of the San Francisco Convention and Visitors Bureau.

Library of Congress Cataloging-in-Publication Data

Gilliam, Harold.
    Weather of the San Francisco Bay Region / Harold Gilliam.—2nd ed.
        p.   cm. (California natural history guides ; v. 63)
    Includes bibliographical references and index.
    ISBN 0-520-22989-4 (alk. paper).—ISBN 0-520-22990-8 (pbk. : alk. paper)
    1. San Francisco Bay Area (Calif.)—Climate.   I. Title.   II. California
natural history guides ; 63.
CQ984.C22 S263   2002   2001027741
551.69794'6—dc21

Manufactured in China

10   09   08   07   06   05   04   03   02
10   9   8   7   6   5   4   3   2   1

CONTENTS

The Golden Gate is the key to the Bay Region climate.

PREFACE TO THE 2002 EDITION

Since the first edition of this book was published in 1962, the science of the weather has made a quantum leap forward, pushing into realms little known at that time. The story of Bay Area weather in 1962 could not take account of global warming, the depletion of the ozone layer, the coastal surge, and the intricate vagaries of El Niño and La Niña, among other things.

Recent satellite technology and advances in computer models of the atmosphere have created new knowledge about the climatic effects of changing ocean currents and temperatures, cloud formation, alternating high- and low-pressure systems aloft, and the cumulative effects of fossil-fuel burning and consequent air pollution on a global scale. This edition takes account of much of that new knowledge insofar as it may affect the local climate.

The terms "weather" and "climate" are often confused. "Weather" refers to what is happening in the air at any given time and place. "Climate" is simply weather averaged over a period of years. Owing to the geographic diversity of the San Francisco Bay Region, which includes both hills and valleys, land and water, there are numerous "microclimates" with special qualities and quantities of wind and rain, heat and cold, and sun and fog at particular times of year. There is also a general Bay Area climate common to all parts of the region. After the original edition of this book was published, I was interviewed by a television weather reporter who asked me how many microclimates were found in the Bay Area. I demurred at setting any specific number, but he persisted. So I impatiently blurted out the first figure that popped into my head, which was the number of feet in a mile: 5,280. That seemed to satisfy him, and it has since occurred to me that the number is as good as any for this purpose. Or pick your own. Although Bay Area climates have many things in common, the differences give the region its extraordinary climatic flavor.

This plurality of climates requires certain departures from the conventional approaches to scientific subjects. Official forecasters cannot make 5,280 Bay Area predictions, and they do not directly measure the microclimates but confine their observations to the regional picture. The varying weathers are experienced only by the residents of the microclimatic neighborhoods, so descriptions of them are principally subjective and anecdotal. Consequently, the story of the Bay Area climates is often based on individual experience. I trust readers will indulge my occasional departures from scientific detachment into personal observation.

A note on style changes for the new edition: Although the weatherman is an almost totemic figure in American lore, we no longer use gender-specific occupational titles—for good reason—but I could not bring myself to use the politically correct term "weatherperson." So I decided to settle for the scientifically accurate title "meteorologist." Of course, you might suppose that a meteorologist would be a person studying meteors. And you would be correct. A meteorologist does study meteors.

However, the meteors in this case are not shooting stars. "Meteor" is from a Greek root meaning something like "phenomenon in the sky." So any natural phenomenon in the atmosphere is a meteor, whether it is a cloud, rain, a rainbow, lightning, snow—or a burning particle from space set on fire by friction with the Earth's atmosphere. Meteorologists are happy to leave those incandescent rocks to astronomers, but they assume jurisdiction over everything else up there as it affects the climate.

Full disclosure department: I am not a meteorologist. I am a writer trying to interpret the recondite language of meteorology for the general reader. I have had to simplify certain technicalities in a way that might not be countenanced by a professor of meteorology if the statement showed up in a student's final exam. But to the best of my knowledge, the facts as stated here are accurate to a first approximation. If there are any errors, they are my own and not those of the experts I have consulted. If you want to look more deeply into weather science, check the sources in the "Further Reading" section at the back of this book.

I am very grateful for the help of several weather professionals whose patient explanations of complicated phenomena have made this book possible, including C. Robert Elford, former state climatologist; meteorologists John Monteverdi and Dave Dempsey of the Department of Geosciences, San Francisco State University; Jan Null, of the same department, formerly lead forecaster with the National Weather Service; Kelly Redmond, regional climatologist at the Western Regional Climate Center in Reno; Orman Granger of the Geography Department at the University of California, Berkeley; and Daniel B. Luten, formerly of the Berkeley Geography Department, my instructor for forty years in matters pertaining to the planet earth.

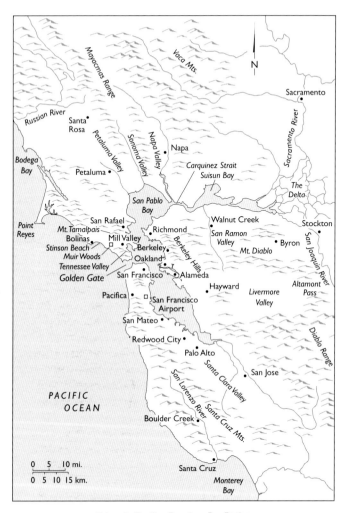

Figure 1. The San Francisco Bay Region

INTRODUCTION

Fair is foul, and foul is fair:
Hover through the fog and filthy air.

Shakespeare, *Macbeth* 1.1

Begin with the most distinctive aspect of the Bay Region climate—the diverse forms of fog. Traditionally, fog of any kind is a grim nuisance that seems to come from nowhere, hides the sun, obscures the terrain, and casts a damp pall over the land. Familiar with the murk that in their times darkened the streets of London, Shakespeare, Dickens, and many another writer associated fog with lurking evil. Combined with smoke, it creates the noxious urban miasma of smog. Dictionaries also identify "foggy" with vagueness, confusion, and other negative qualities.

In the San Francisco Bay Region, however, the characteristic summer fog is widely regarded as a boon and a blessing, bringing the clean, salty aroma of the Pacific, keeping the days refreshingly cool, advancing from the ocean in fantastic flowing forms that from day to day are unique and unpredictable. Shaped by the slant of the California coastline and the oceanic currents of wind and water, by the presence of a strait and estuary penetrating the coastal mountain ranges, by the hill-and-valley contours of San Francisco and the region around the bay, the advance of the summer fog inland from the Pacific is one of the planet's most awesome natural spectacles.

My own first experience with Bay Region fog occurred when I was a teenager traveling with my family from southern California to San Francisco. We were on the old Skyline Highway along the spine of the Santa Cruz Mountains when the road emerged from forests into an open meadow, affording a panorama that made us stop in amazement. Below us, moving in from the ocean, was a sea of billowing vapors, dazzling white in the late afternoon sun.

My father pulled the car off the road to an area where other travelers had stopped to gaze at the spectacle. Rising into the clear, dry air above the vapory sea were separate peaks and ridges, islands in the white flood. Waves of fog were moving eastward below us like ocean breakers lapping on the shores of an archipelago. We watched until the sun was at the horizon, throwing into sharp relief the shadowed troughs of the waves below their gleaming white crests.

Following the highway northward, we descended into that sea of fog and for several miles could scarcely see beyond the hood of the car. When we arrived in San Francisco, the fog was overhead, and it seemed as if we were in a city at the bottom of the ocean.

In later years, living in the city, I observed an endless variety of fog phenomena that had never been scientifically catalogued. On summer afternoons, emerging from work at the Chronicle building on Mission Street in downtown San Francisco, I could look west and see a white cascade of oceanic vapor pouring over Twin Peaks and flowing down the near slope like some slow-motion Niagara. My way home was in that direction, and en route I encountered advancing fog masses several hundred feet high, moving not only over Twin Peaks but through the low gaps in the city's hilly terrain, particularly the valley of Golden Gate Park. By the time I reached home near Buena Vista Park, I was surrounded by the thick rolling vapors, although downtown was still in bright sunshine. Wherever I have lived in San Francisco—the Presidio, Buena Vista, Pacific Heights, Telegraph Hill, Cole Valley, Sunset Heights—I have observed that the weather in my neighborhood, particularly in the summer, was often drastically different from the atmospheric conditions a few blocks away, owing to the city's hill-and-valley topography. As the fog advances with the breeze from the ocean, it envelops streets and houses on the windward slopes and leaves the sheltered leeward valleys warm and sunny. If the white tide from the Pacific continues to advance, it will surmount the hills, flow down the leeward slopes, fill the valleys, and advance to the next hill, and perhaps beyond to the next valley or to the bay.

The picture is similar on a larger scale throughout the nine-county Bay Area. Fishermen along the fog-shrouded coast of Marin County on a summer day may be shivering in the low fifties while people in San Rafael, ten miles east, bask in comfortable 70-degree weather and residents of ranches at the edge of the Sacramento Valley, another forty miles east, mop their brows as the thermometer hits 100—a temperature difference of 50 degrees in fifty miles.

During the winter there are similar differences. Ben Lomond, in the Santa Cruz Mountains, averages 42 inches of rain a year, while just over the hills, in the Santa Clara Valley, the annual total is only 13 inches.

We have not yet fulfilled the age-old dream of controlling the weather, but in the Bay Region we come close; we can change the weather around us by moving a short distance. Probably no comparable area on earth displays as many varieties of weather simultaneously as the region around San Francisco Bay.

THE WEATHER FUNNEL

The reasons for this unique situation lie in California's extraordinary geography. In general, the state has a Mediterranean climate, with mild, wet winters and dry summers. But that general type is locally modified by special features of the landscape. The prime movers in setting up the geography were two mountain ranges, a river system, and a big thaw.

The Sierra Nevada, a granite barrier rising from 10,000 to

14,000 feet into the sky some two hundred miles inland from the shore of the Pacific, intercepts the clouds and moisture-laden winds moving eastward from the ocean and forces them to drop their burdens on the mountain slopes in the form of rain and snow.

The water, cascading down the western slope of the Sierra in an intricate network of creeks, waterfalls, streams, and rivers, merges in the Central Valley to form the greatest river system within the boundaries of any single state. This tremendous volume of water, slicing through the Coast Range to the sea, carved the Carquinez Strait and the Golden Gate long before San Francisco Bay was formed.

At the end of the last Ice Age, the great glaciers melted in such volume that the oceans overflowed. Over a period of thousands of years, rising seas flooded through the river-carved gorge at the Golden Gate and occupied an inland valley to create San Francisco Bay. The river no longer flowed to the sea but emptied into the bay at the Carquinez Strait.

Thus the successive action of the river and the ocean created the only complete breach in the Coast Range, which borders the Pacific for most of California's length. As a result, the Bay Region is the meeting place of continental and oceanic air masses. Through the funnel of the Golden Gate and San Francisco Bay, the immense aerial forces of sea and land wage a continual war, and the tide of battle often flows back and forth with regularity. The line between the two types of air masses, particularly in summer, may zigzag through the streets of San Francisco and extend in similar erratic fashion across the entire region.

THE SUBDIVIDED RANGE

The reason for the zigzags—the highly variable weather patterns within the region—is the complex topography of the Coast Range, which modifies the basic struggle between air masses of land and sea in intricate ways. (Although the Coast Range is actually a series of ranges, it is more convenient to refer to it here in the singular.) The weather of any mountain range, with its ridges and canyons and valleys, is complex enough, but the section of the Coast Range comprising the Bay Region divides and subdivides into various subranges, each with its own hill-and-valley contours, creating its own modifications of the basic weather and climate patterns.

In general, the Coast Range in this region is a double chain of mountains running north and south (or, more precisely, north-northwest and south-southeast). Between the two chains lies the basin of San Francisco Bay, including the valleys at the ends of the bay: Petaluma on the north and Santa Clara on the south (see fig. 1, facing p. 1).

The western range consists of the Santa Cruz Mountains, south of the Golden Gate, and the Marin hills, including Mount Tamalpais, to the north. As if to complicate matters further, the eastern part of the Coast Range in this vicinity is itself divided into two main chains. Immediately to the east of the bay are the Berkeley Hills, paralleled, beyond the San Ramon and Livermore valleys, by the higher Diablo Range. North of the bay, this double aspect of the range continues, but it is further subdivided into subsidiary chains, including the Sonoma, the Mayacmas, and the Vaca mountains, which separate the legendary wine-grape valleys of Napa and Sonoma.

MICROCLIMATES

Eastward from the ocean, over the several ranges, each successive valley has less of a damp, seacoast climate and more of a dry, continental climate—hotter in summer and colder in winter. But this basic pattern is further modified and complicated by a number of gaps and passes in the ranges—the most important of which is the Golden Gate—that allow the easy penetration of seacoast weather inland.

The pattern is also modified by large bodies of water, which tend to cool their shores in the summer and warm them in the winter. The most important of these is, of course, San Francisco Bay itself and its various subdivisions and tributaries, including San Pablo Bay, Suisun Bay, and the Sacramento–San Joaquin Delta, where the major rivers of the Sierra and the Central Valley converge in a complex network of water-courses and low islands.

Because the land takes these varied forms, there is actually no such thing as a Bay Region climate. There are only innumerable microclimates within the region, varying widely from mountain to mountain, from valley to valley, and from point to point within the mountains and valleys.

The results are manifold: the great flowing fogs that move through the Golden Gate and over the coastal hills from the ocean in summer; the warm, dry winds that sometimes whip down through the canyons to the bay in spring and fall, scuffing the bay surface into whitecaps; cumulus clouds, which drift eastward across the sky in winter, throwing moving patterns of light and shade across land and water; massive cumulonimbus clouds, which may build up to heights of 50,000 feet or more above the rim of mountains around the bay; rains that deluge one valley while scarcely dampening the next; snows that occasionally dust the tops of the highest peaks; and the rare frosts that whiten the lowlands.

All these phenomena are of minor concern to the average city dweller, who merely has to decide whether to take a coat or umbrella to the office in the morning. But they are of great interest to those who are directly affected by the weather, including pilots of ships and small boats, who must learn to navigate in the fog and cope with churning seas stirred up by winds; crews who repair power lines and telephone wires damaged by storms; road workers who keep the streets and highways clear of debris in heavy rains and winds; the passengers and crews of the hundreds of planes whose schedules may be upset by blinding fog; carpenters and construction workers, who cannot operate in storms; dairymen, who must spend extra money for hay if not enough rain falls to raise a good crop of grass; painters on the bay's great bridges; grape growers in Napa and Sonoma and Livermore, who fear late frosts and early rains; and farmers in the interior valleys, who must reckon when to prune and irrigate and plow by observing signs in the sky.

For the seven million people who live and work in the San Francisco Bay Area, the weather is the lowest common denominator, consistently the most recurrent topic of conversation on street corners, in corridors and elevators, in taverns, on buses and trains, and wherever people gather to talk. Although weather talk is common worldwide, here it achieves a particular flavor and intensity unknown elsewhere.

Residents of other parts of the country are likely to open a conversation with a remark such as: "Some weather we're having lately, isn't it?" But in the Bay Region no one—except tourists and newcomers—assumes that the listener has been experiencing the same kind of weather as the speaker, and the opening gambit is likely to be: "What's the weather like where you live?" Commuters to San Francisco compare notes on their respective communities, and no one is greatly surprised if Berkeley has fog while Alameda is in bright sun, or if Mill Valley has rain while Palo Alto is dry.

THE OCEAN OF AIR

Before getting into the detailed peculiarities of the Bay Region's microclimates, we should review some general facts about the atmosphere that apply to all weather and climate everywhere on earth.

1. The ocean of atmosphere that surrounds the earth bears down on the earth's surface with an average "weight," or pressure, of 14.7 pounds per square inch at sea level. The exact pressure depends, however, on the altitude, air temperature, and other variables.
2. Warm air is light and tends to rise; cold air is heavy and descends. Because the cold air presses down more heavily on the earth's surface than warm air, cold air causes high pressure; warm air, low pressure. (Temperature is not the only determinant

Rotate paper to left, counterclockwise (direction of earth's rotation as seen from above the North Pole)

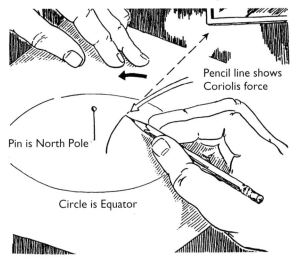

Pencil line shows Coriolis force

Pin is North Pole

Circle is Equator

Figure 2. The Coriolis force

of high and low pressure, but it is a key to the formation of San Francisco's distinctive fogs.)

3. Just as water tends to seek its own level, so air tends to equalize its pressure. Thus, air moves horizontally from a high-pressure area to a low-pressure area; winds blow from a cool to a warm zone, just as cold air will move through a door or window into a warm room, creating a draft. (There are exceptions to this, too.)

4. When dry air rises, in general it expands and cools at the rate of about 5.5 degrees Fahrenheit per thousand feet of elevation.

5. When air descends, it compresses and grows warmer at about the same rate.

6. Warm air is able to hold more moisture (in the form of vapor, which is invisible) than cool air. If warm,

damp air begins to cool off, it will reach a point where it can no longer contain its vapory load (the dew point), and the moisture will condense into drops of visible fog or cloud. A familiar example is the condensation that occurs when air is cooled by contact with a glass of cold water and the condensed moisture is deposited on the outside of the glass.

7. Air moving freely across the surface of the earth tends to curve to the right, clockwise, in the Northern Hemisphere, and to the left in the Southern Hemisphere. This tendency is a result of the Coriolis force, or Coriolis effect—named for the French scientist who formulated the principle.

To illustrate the Coriolis force, draw a large circle with a dot in the center on a piece of paper (see fig. 2). The circle represents the earth as seen from above the North Pole, and the dot is the pole itself. Rotate the paper slowly counterclockwise, representing the rotation of the earth. As the paper rotates, draw a short line inside the circle from any point toward some fixed point off the page, such as the wall of the room. You will find that the line on the paper curves to the right.

In the same fashion, anything moving freely over the surface of the Northern Hemisphere tends to curve to the right because of the rotation of the earth beneath it. (In the Southern Hemisphere, the curve is to the left.) This is true not only of winds but of ocean currents, rockets, and artillery projectiles. Even rifles must be compensated to counteract the drift of the bullet. This clockwise motion of winds and ocean currents is particularly important to understanding the Bay Region weather.

THE FOUR SEASONS

SPRING

During his stay in San Francisco in the 1880s, Robert Louis Stevenson wrote of the "trade winds" that blow in through the Golden Gate and over the hills of the city. The term conveys the proper romantic flavor and conjures visions of fabled galleons on the trade routes to Cathay, but the fact is that no trade winds blow within thousands of miles of San Francisco. Countless writers, attempting to lend a note of glamour to the prevailing sea breeze that blows from the ocean across the city's hills, have repeated Stevenson's error.

The trade winds blow from the east in the latitudes on either side of the Equator. Spanish galleons of the sixteenth and seventeenth centuries made use of the Pacific trades. They sailed south from Acapulco, Mexico, to take advantage of these winds, which bore them west across the Pacific to Manila. On the return trip, however, they sailed far to the north in order to pick up the prevailing westerly winds, which blew from the west across the ocean in the latitude of the United States. The westerlies brought them to the coast of California; from there they could sail south to Mexico.

It is the westerlies that assault the California coast, throwing sheets of ocean spray against the headlands, blowing sand along the beaches into great dunes and drifts, flailing the grasses of the coastal slopes, bending the laurels and cypresses into permanently contorted shapes.

In the spring, the westerlies are greatly intensified. During March, April, and May, as the days grow longer, the northward-moving sun heats the continent day by day, melting mountain snows, causing flowers to bloom and fruit to bud across foothills and plains and deserts. In the Central Valley of California, a 500-mile-long basin surrounded by mountains, the heated air rises. Through a complicated interchange of pressure systems in the upper atmosphere, the result is falling pressure at ground level in the Valley, recorded on barometers there. Weather Service forecasters mark the Valley on their maps with a large "L" for low pressure. Out over the Pacific, however, the air in contact with the ocean is cool and heavy. It presses down on the water surface; ships and weather buoys relay the barometer readings to the Weather Service, and the forecasters indicate the high pressure on their maps with the letter "H." The result of this differing pressure is what the meteorologists call an "onshore pressure gradient." The air rushes from the high-pressure areas toward the low, from the cool ocean toward the warming continent.

In March and April, the weather in the Bay Region is highly variable. Days marked by the last gusts of winter blowing themselves out are followed by days of hazy, balmy warmth. The green hills, the grassy vacant lots, and the residents of the region absorb the increasing sunlight. A light haze hangs over the bay and the hills as moisture is evaporated into the air from the water and from the earth, often still damp from the last showers of winter. Along the ocean boundaries of the land, the battering waves of winter die out and are replaced by gentle swells.

The surface of San Francisco Bay is glassy. By many an afternoon in early spring, the sun has sufficiently heated the land around the bay to cause the surface air to warm and rise. Cooler air moves in from the ocean, ruffles the bay, and clears away the haze.

Meanwhile, as if with a far-off roll of drums, the forces of the atmosphere are being marshaled for a major change.

The Pacific High

Air warmed by the hot sun over the Equator rises and heads northward toward the Arctic. Some of it cools off and sinks to the ocean surface several thousand miles to the north, where it becomes the Pacific High—a "mountain" of cool air weighing heavily on the water.

Marked by a big "H" on the weather maps, the Pacific High usually occupies a position somewhere between San Francisco and the Hawaiian Islands, sending out winds in all directions across the ocean surface. Because it is a result of the sun's heating, it usually follows the sun to the south in the winter and migrates northward with the sun in the spring. (Of course, the sun does not literally move north. It appears to do so as the planet swings around to the "springtime" position in its orbit and the tilt of the earth's axis exposes the Northern Hemisphere more directly to the sun's rays.)

Like everything else moving on the surface of the earth, the winds coming from the Pacific High toward the continent are affected by the Coriolis force; they begin to curve to the right. When they reach California, they are coming from the northwest, in some places moving almost parallel to the slanting coastline.

There they are prevented from moving inland in force by the high wall of the Coast Range, and so they blow on down the coast, breaching the mountain barrier only where gaps and passes in the range permit tongues of ocean air to penetrate inland.

Ocean River

Through March and April and May, the Pacific High moves closer and grows stronger, while the Central Valley continues to warm up. At any latitude, ocean currents are created and borne along by the action of the wind. The northwest wind

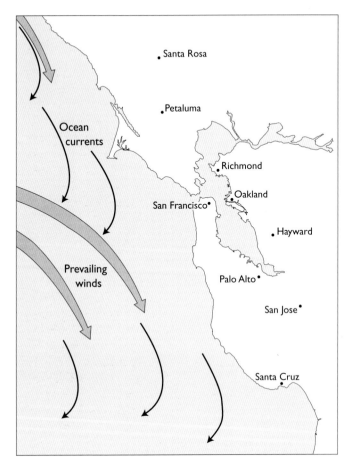

Figure 3. Offshore upwelling

blowing down the California coast pushes the surface of the ocean before it and creates a strong current running southward down the shoreline like a river. But again the picture is changed by the Coriolis force. Like the winds themselves, the waters, too, tend to curve to the right across the surface of the rotating sphere. The southward-moving current veers offshore. In order to replace the surface currents moving away from the coast, masses of water surge up from the bottom of the ocean, creating a continual fountain of upwelling waters (see fig. 3).

This bottom water, which may come from depths of several hundred feet, may be 10 to 15 degrees colder than the sun-warmed sea surface. Summer swimmers at northern California beaches are painfully aware that these waters are frigid—often in the mid 50s—compared to the temperatures of the surf along the beaches of southern California, which are beyond the principal zone of upwelling and frequently reach the 70s. In some areas where the upwelling is intense, the water in summer is colder than in winter, when the upwelling is minimal.

The Great Fog Bank

This streak of cold water along the coast is a basic part of the Bay Region's summertime air-conditioning system. The wind from the northwest, skimming thousands of miles of ocean, absorbs great quantities of moisture that has evaporated from the surface. The moisture is suspended in the moving air as invisible vapor. When it approaches the coast, the air comes into contact with the cold, upwelled waters and is itself cooled off, causing its invisible vapor to condense into visible droplets. The droplets cling to the minute particles of salt that have been thrown into the air with the spray. As the wind blows over the cold surface, the drops continue to form until they create a haze, which soon thickens into fog. The same process, explained earlier, that causes moisture to form on the outside of a glass of cold water takes place here on a mammoth scale.

The result is the great fog bank that envelops most of the California coast intermittently during the late spring and the summer. It may range in width from a hundred yards to more than a hundred miles, in height from a hundred feet to half a mile. In the beginning, however, it takes form merely as puffs and wisps of vapor that drift landward and cling to the shoreward slopes of the coastal hills or hang over the beaches. Then, as the season advances and the wind increases, the entire process accelerates; the separate vapors merge into a solid mass that steals up the coastal canyons in late afternoon or early evening when the sun's rays are no longer strong enough to burn it off. In some of the canyons, it collects on the branches of the redwoods and drips to the ground, keeping the earth damp. The big trees thrive on this moisture. Redwood country is fog country; *Sequoia sempervirens* rarely grows naturally beyond the range of the sea fog.

Along most of the Coast Range, the sea air and its fog reach the heads of the canyons and are stopped by the higher ridges from penetrating farther inland. But at the Golden Gate, the only sea-level breach in the mountains, the wind moves through the range, bringing with it the masses of condensed moisture. The fog may be first visible in the Golden Gate in early spring, perhaps appearing on an April morning as a wispy finger of white moisture entering the bay beneath the 230-foot-high deck of the Golden Gate Bridge, heralded by the sonorous trumpeting of the fog horns. In the course of a few hours, it may thicken until it is a solid mass moving through the milewide strait into the bay. At the maximum, an estimated million tons of water an hour float through the Gate as vapor and fog.

The moving mass of visible moisture envelops the deck of the bridge and swallows ships and shores as it advances. But the fog itself is not the primary element at work. It is created by and borne upon the wind. When the wind is sufficiently laden with moisture and sufficiently cooled by the water, the fog comes into being and is carried inland by the moving air. Thus the wind strikes a particular area in advance of the fog. (Both wind and fog funneling through the Golden Gate created highly dangerous conditions for workers during the construction of the bridge in the 1930s.)

The fog formed in this way is literally a low cloud. It may later rise into the air and look more like the conventional idea of a cloud, but it does not change except in altitude. Visitors are often puzzled when Californians refer to this cloud layer as a "high fog," but the terms are interchangeable. Meteorologists call it "stratus," which is one type of cloud (see p. 52), whether it is at ground level or overhead.

Fog Fantasies

Fantastic fog forms may develop as the advancing white mass encounters obstacles. It may come in surges like a slow-motion surf, exploding into spray on the ridge at the north end of the Golden Gate Bridge, forming a standing wave over Sausalito, whirling horizontally in eddies around promontories, and pouring over Twin Peaks and the Peninsula hills, where it forms fog falls and fog cascades down the leeward slopes. If it comes in low on the bay surface, it is likely to billow in domes over Alcatraz and Angel islands. At times a fog deck will appear part way up the Berkeley Hills and build out toward the bay.

Sometimes, fog is formed suddenly by very odd circumstances. When the damp ocean air is almost at the saturation point, but condensation has not yet occurred, anything that lowers the temperature a few degrees can cause condensation to take place. A high layer of clouds moving in front of the sun and casting a shadow can cause sufficient cooling of the air for stratus to form. An eclipse of the sun has been known to create a sudden fog lasting for the duration of the eclipse. Nearly saturated air moving across a small body of water, such as Stow Lake in Golden Gate Park, will drop in temperature sufficiently on contacting the slightly cooler water to form small wisps of fog, which drift across the lake, duplicating in miniature the stratus formations over the ocean.

A similar phenomenon on a larger scale takes place in the Golden Gate when clear air near the saturation point is cooled a degree or two by a reversal of the tidal current beneath it. When the tide changes from ebb to flood, the warm outgoing waters of the bay and its tributary rivers are replaced by

Figure 4. A plane leaves a contrail made up of
condensed vapor from the engine exhaust.

the cold incoming waters of the ocean. The cold water lowers the air temperature enough for the vapor to condense into
fog. Similarly, a tidal shift from flood to ebb may cause the
fog to disappear.

The spring and summer fogs of the Bay Region are known
as "advection" fogs. Just as "convection" refers to a vertical
movement of air or liquid (as takes place when water boils),
advection is a horizontal movement, and an advection fog is
formed by the horizontal movement of the air.

When the ocean wind is not damp enough or the water cold
enough to form stratus over the surface, the formation may
take place when the wind hits the coastal hills and is forced to
rise. As it rises, it cools sufficiently to cause its vapor to condense in a fog deck against the hills, or perhaps in isolated puffs
and wisps, which are then carried inland by the wind and float
near rooftop level across downtown San Francisco.

Another instance of condensation occurs when a plane at
a high altitude creates its own streak of "fog" by the emission
of water vapor from its exhaust (see fig. 4). Whether condensation trails (contrails) develop depends on the temperature and moisture content of the ambient air.

Plate 1. Fingers of summer fog flow through the Golden Gate into San Francisco Bay. This kind of fog, which forms as a relatively flat layer, is known as "stratus."

Plate 2. On the mile-long span of the Golden Gate Bridge, drivers on a summer day may encounter both sun and fog.

Plate 3. As the afternoon advances and the heat of the sun diminishes, the stratus layer thickens.

Plate 4. The north tower of the Golden Gate Bridge frames San Francisco's Transamerica spire above the fog blanket.

Plate 5. A wave of summer stratus advances from the ocean and rises like a breaker over the hill at the north end of the Golden Gate Bridge.

Plate 6. As the flow of fog from the ocean continues over a period of hours, it lifts from the deck of the bridge and will eventually overtop the bridge's 750-foot towers.

Plate 7. During the construction of the Golden Gate Bridge in the 1930s, reduced visibility and strong winds were high-risk hazards for workers on the bridge's cables.

Plate 8. A stratus layer from the ocean flows over the city's central hills, forming a fog cascade down the eastern slopes of Twin Peaks.

Plates 9 and 10. A fog wave forms over the ridge behind Sausalito on a summer afternoon . . .

. . . and descends like a curtain over the town in the evening.

Plate 11. As early-morning sunlight strikes the western districts of San Francisco, the overnight stratus layer begins to burn off from the top, leaving residual pools of fog in low areas. The pools will soon evaporate as the sun reaches them, but over the cold waters of the Golden Gate (beyond), the fog will remain longer, to diminish and disappear when the sun is high.

Plate 12. Below Mount Davidson (seen here looking south from Twin Peaks), fog pools evaporate as the morning sun penetrates a cloud ceiling.

Plate 13. At the Golden Gate, the stratus is slowly burned off as the warming air dissolves the moisture, clearing the deck. Soon the towers will emerge into the sunlight.

Plates 14 and 15. Sunbathers at Baker Beach and pilots guiding ships through the Golden Gate are happy about the burnoff . . .

. . . but offshore from Baker Beach, a massive bank of stratus hovers, ready to advance again when the sun sinks in the west.

SUMMER

Week by week through March and April and into June, the forces that produce the stratus increase in intensity. The Pacific High moves farther north, closer to the latitude of San Francisco, sending out stronger winds; offshore the upwelling of cold bottom waters increases, condensing the winds' moisture into thicker masses of fog; in the Central Valley, the northward-moving sun sends temperatures to the 100 mark and beyond.

Along the Valley floor, in orchards, vineyards, and cotton fields, the crops ripen in the heat. The hot air rises, sucking cool masses in great drafts through the only break in the Valley's surrounding mountains, San Francisco Bay. With the ocean air comes the fog, evaporating gradually in the hot, dry air of the Valley, but sometimes penetrating at night as far as Sacramento and Stockton.

Visitors and recent arrivals to the Bay Region are often puzzled by the sharp changes in the summer weather from day to day. Some days the bay and its shores are totally fog-bound, but other days are clear and cool. The fog seems to come and go in cycles. Until recent years, the conventional explanation for the fog's behavior was a simple one: as the cool, fog-bearing ocean air is pulled over the coastal hills and across the bay toward the hot Central Valley (i.e., from a high-pressure area to a low), the nearest parts of the Valley begin to cool off after a few days, much as a draft from an open door lowers the temperature in a warm room.

The incoming cool, heavy sea air replaces the rising warm land air, and temperatures in Sacramento and Stockton may drop from well above 100 degrees to the "cool" 90s. When the Valley cools sufficiently, the fog-producing machinery breaks down. Without the intense Valley heat to draw the sea air in through the Bay Region, the wind diminishes and no longer carries the fog inland. San Francisco, the Golden Gate, and the coastline are fog free.

Then the process starts all over again. Without the incoming wind and fog, the sun gradually reheats the Valley. The rising warm air again begins to attract the foggy marine air inland, through the Golden Gate and across the bay. The result is a fog cycle of about a week in length, producing roughly three or four days of fog over the bay and three or four days of sun.

One problem with this traditional explanation of the summer regime is that the foggy or sunny spells sometimes persist for weeks, with no apparent cyclical regularity. With increasing knowledge of the workings of the upper atmosphere, meteorologists have learned that although the fog penetration is related to the alternate heating and cooling of the Valley, it is also influenced by much larger forces in the global weather picture.

The Jet Stream

Until the middle of the twentieth century, weather observations were primarily confined to measurements of events on the land and sea surfaces. Then, during World War II, pilots flying at previously unattained heights of 30,000 feet and above discovered, at the core of the prevailing westerlies, a jet stream—a river of air moving at speeds of 150 miles per hour or more. This powerful current, which flows from west to east over the Pacific and across North America, is also known as the "storm track," because it carries with it storms that bring rain and snow to the continent and affects local weather in ways that were previously unexplained.

It travels not in a straight thrust but in a weaving pattern that loops north and south in unpredictable S-curves. Two major jet streams circle the globe in the Northern Hemisphere, and two in the Southern Hemisphere, but the one primarily affecting the United States is the polar jet stream, so called not because it moves over the pole but because it is the farthest north of the four. The latitude of this storm track depends on the season. Its most southerly course, in the winter, may cross the southern United States. Like the Pacific High, it moves northward with the sun in the spring and summer, and southward in the fall and winter. Its normal summer track is across central Canada, but it brings rain to a wide swath of the continent, often drenching Vancouver, Seattle, and Portland.

Although California is south of the summer rainfall pattern, the remote influence of the jet stream is felt even in the Bay Region. Associated with it are recurrent waves of high-altitude pressure changes. North-south "ridges" (high pressure) and "troughs" (low pressure) advance from west to east, over the ocean and onto the continent, much like successive crests and troughs of waves on the ocean.

Ridges are characterized by clear weather, troughs by storms. A trough bringing rain to Canada and the Northwest can influence weather for thousands of miles, as far south as the latitude of the Bay Region. A weak southern extension of that trough, even though it may be at an altitude of 30,000 feet or more, can affect the pressure on the surface of the sea or land beneath it. The low pressure of the trough over the ocean can weaken the Pacific High, diminishing the onshore winds it sends to the California coast. As the winds slacken, the upwelling of cold water from the sea bottom diminishes or stops; and the fog-making process slows to a halt. The result is clear weather along the coast and sunshine in the Bay Region, occasionally accompanied by high, thin cloudiness associated with the distant storms passing over Canada.

By the time the eastward-moving trough of low pressure reaches California, it reinforces the heat-generated low over the Central Valley, attracting the cool marine air and its fog. Simultaneously, out over the ocean, it has been succeeded by a ridge, reinvigorating the Pacific High, which again sends its winds to the coast, resuming the upwelling. Fog over the bay again results from the push of the Pacific High and the pull of the Valley heat.

In this way, the ridge-and-trough pattern acts to strengthen or weaken the weekly cycle that results from the heating and cooling of the Valley. It also explains why the normal cycles may be interrupted by spells of sun or fog persisting for much longer periods. In some summers, there scarcely seems to be any sun at all; other summers may provide ample beach weather.

The explanation for these anomalies lies in the complicated behavior of the jet stream and its associated pressure changes. At times, the successive west-to-east waves of high and low pressure aloft—unlike waves on the ocean—will slow down and stop, owing in part to changes in the temperature of the ocean surface. A ridge or a trough will remain overhead for weeks, resulting in extended spells of clear or foggy weather. Long periods of fog on the bay, for example, will usually coincide with persistent downpours of unending rain in Vancouver and Seattle, to the frustration of residents of both regions.

Multicycles

There are also long-term influences on fog formation associated with the El Niño phenomenon, to be explained later. But aside from these influences and the occasional pauses in the eastward movement of the ridges and troughs, an approximate weekly cycle still prevails. There are other cycles as well, including one that takes place daily. During the foggy part of the weekly cycle, in late afternoon, as the sun's radiation weakens, the fog moves in through the Golden Gate and other gaps in the coastal hills, spreads over a large area of the bay and its shores during the cool hours of the night, then is burned off by the morning sun. It disappears from the top down, sometimes leaving in low areas residual pools that soon evaporate.

The daily cycle takes place in this particular way only at the early stages of the weekly cycle, when there is an equilibrium of the fog-producing forces. Later, when the stratus covers the coast and the central Bay Region for days at a time, the daily cycle is more evident farther inland. San Francisco and the immediate vicinity may be blanketed for several successive days, while Palo Alto to the south and San Rafael to the north are beyond the fog's reach. At night, however, the white mists may roll in over the coastal hills, engulfing everything from the Santa Clara Valley to the Napa Valley and as far inland as Stockton, only to be burned off in these outer areas by the morning sun.

When the sun reappears after a spell of fog on the coast, residents make the most of the transient warmth; they know from experience that the great fog bank will soon return. Usually, it will first become visible offshore, flowing down from the northwest as the prevailing winds resume, but on some occasions, particularly in the fall, after a period of warm easterly winds, a special combination of circumstances will bring the fog from southern California in a phenomenon known as a "coastal surge." At such times, fog developing around Point Conception, near Santa Barbara, moves northward. In about two days, it may reach Monterey, and it is likely to blanket the coast northward overnight and to be flowing through the Golden Gate by the third day.

The third type of cycle is seasonal; it begins in early spring. As the Pacific High grows stronger and the Valley heat increases, the stratus penetrates farther inland and remains longer with each weekly cycle, reaching a maximum in August, then decreasing through September.

Finally, there are long-term tendencies perhaps too irregular to be called cycles. Several successive summers with very little fog may be followed by several very foggy summers. On the other hand, a foggy summer may be succeeded the next year by a very clear summer. The reasons are related to the El Niño phenomenon and will be described later (see p. 91).

The Inversion

Although the advection fog is formed primarily by the horizontal movement of air, in the development of the fog formations and cycles, vertical movements are important as well. As the cool, heavy ocean air moves in across the bay and its shores, it slides beneath the warmer air. Thus there is a layer of cool, moist air at the surface, topped by warmer air above.

Because the more normal condition of the atmosphere around the world is that air is warm at the surface and becomes cooler at higher elevations, this situation over the Bay Region (and over much of the Pacific Coast) is an "inversion"—the usual atmospheric condition is inverted.

Anyone who climbs the hills around the bay in midsummer may be aware of sharp changes in temperature that come with changes in elevation. At a certain height above the bay, a climber will move from the cool surface layer into the warm air above, actually climbing into the inversion. The temperature may suddenly rise 15 degrees within the rise of 20 or 30 feet in elevation.

This warm layer acts as a lid or ceiling above the cool ocean air and its fog. When the stratus comes in, it will rise to this inversion layer and stop, unable to penetrate the invisible barrier of warm air.

As the cool air flows in from the ocean with increasing volume beneath the inversion, it is able, literally, to raise the lid, forcing the warm air layer upward. Thus, the fog roof rises from an elevation of possibly 200 feet to a normal maximum of about 2,000 feet. In the early stages of the daily or weekly cycle, the height of the inversion layer can often be estimated by a glance at the Golden Gate Bridge. In the spring, the inversion lid may initially be about at the level of the 230-foot-high deck of the bridge, and the stratus flows underneath, while the traffic moves through the warm, clear air of the inversion layer itself. Over a period ranging from hours to days, the fog roof rises higher, until it engulfs even the tops of the 750-foot-high towers, pushing the inversion layer up to that elevation.

High Fog

At about the same time, another phenomenon occurs, which changes the entire aspect of the stratus on the bay. As the fog mass moves eastward, it crosses surfaces warmer than the frigid upwelled waters of the ocean. One such surface is the sun-heated water of the bay, where very little upwelling takes place. Another is San Francisco itself, and other land to the north and south. When the fog-bearing layer moves over these areas, it is heated a few degrees; as the air warms, the moisture begins to evaporate, leaving the lower air clear. The fog appears to lift off the bay and the city. It may remain at street level in the western districts of the city, but it gradually warms and evaporates as it moves east, and by the time it reaches the downtown area, it will be some distance overhead.

The higher air, not warmed by direct contact with the land or the bay, remains cool; and its moisture does not evaporate. The result is a layer of clear air below a gray ceiling—a "high fog." The Golden Gate comes slowly into view again, as if a misty curtain were rising, revealing the bridge from the bottom upward. The ceiling slants up to the east, and the incoming cool air piles up against the Berkeley Hills.

So the fog does not flow through the streets of flatland Berkeley but envelops the hill districts at various elevations, depending on the depth of the stratus layer and the force of the breeze bringing it from the ocean. There is no consistent fog level in Berkeley's hill neighborhoods. Homes at any location may be above or below the fog deck at any given time—or completely immersed.

As the low surface layer of cool air is replenished from the ocean and increases in depth over the bay and its shores, the temperatures in the region near the Golden Gate continually drop toward the temperature of the ocean—the low 50s. San Francisco's coldest spring and summer weather comes, not when the stratus is low in the streets, but later, when it has been pushed high overhead.

During the summer, the inversion, once established, is relatively permanent. It rises and falls with each cycle, but seldom gets as low as in early spring (or in the autumn), when the fog in the Golden Gate may be close to the water. When the summer stratus returns after a few clear days, it will usually form, not on the water, but at some distance overhead, where the layer of ocean air meets the inversion. Such high fog is typical of midsummer. When the stratus is still on the ground (at least in hilly areas), it is possible to distinguish between two varieties of fog, the wet and the dry. Both types may roll through the streets in almost impenetrable billows. Wet fog (sometimes called "Oregon mist") collects on trees and wires, forming large drops, which fall to the ground. Automobile tires swish along the damp pavements, and windshield wipers click steadily, barely able to keep the glass clear. Dry fog may be equally dense but does not produce a drip. Even though a driver may scarcely be able to see beyond the hood of his car, no moisture collects on the windshield.

The wetness or dryness of a fog may be the result of various conditions, among them the distance the fog has traveled and the depth of the stratus layer. If the fog-forming process has begun far out at sea, the drops of moisture have ample time to grow and merge as the fog-bearing wind moves landward. By the time the fog reaches the coast, it has developed a depth of several hundred feet, and the drops have become large enough to produce a steady drip from leaves, wires, and eaves. The drip will be accelerated if the wind is strong, causing the drops to whirl and collide. A dry fog, however, is formed nearby, perhaps just offshore or above the coastal hills, and the droplets have not had the opportunity to increase in size.

Because of the inversion lid of warm air above it, the foggy layer usually cannot rise and cool sufficiently to turn into rain, but occasionally a wet fog will be thick enough to be measurable in rain gauges. Even when the fog is not thick enough or wet enough to deposit moisture in a rain gauge or on open terrain, fog drip beneath trees may amount to rainfall. A stand of trees in a foggy area will not only collect drops from fog blowing directly into the trees, it will tend to stir up the foggy air above them, causing turbulence, as a rocky stream bed stirs up the water above it. The turbulence sets up vertical eddies, and the air eddying down into the treetops deposits moisture on the leaves and branches.

It is the steady drip from this deposit that provides water

to nurture the redwood groves in the coastal canyons. Another place of copious fog drips is opposite the Golden Gate in the Berkeley Hills, where the fog deck frequently forms on the upper slopes. Eucalyptus and pine groves planted there long ago intercept large amounts of fog and cause a rainlike deposit of moisture. The fog drip there during the summer months has been measured at a surprising ten inches, an amount nearly half as great as the total annual rainfall in Berkeley.

It has been speculated that planting such groves of trees along barren coastal mountains might increase the annual precipitation and add to the water supply. Careful studies would have to be conducted, however, to determine whether such plantings might decrease the fog canopy and the fog drip on areas farther inland.

The fog drip is normally the only kind of precipitation that reaches San Francisco in the summertime. The jet stream and its copious rainstorms are far to the north. The rare summer shower that strikes the Bay Region, perhaps accompanied by thunder and lightning, usually comes in from the south as an errant mass of warm, moist air moving north from the ocean off Mexico. Summer thundershowers in the Sierra are usually caused by the same kind of humid air from the south, rising and cooling on the undulating western slopes of the range. Just as a trough associated with jet-stream storms in the Northwest will tend to bring fog to the Bay Region, so a subsequent ridge (high pressure) is likely to bring clear weather here. Because air blows outward from a high-pressure zone, a particularly strong high over the Northwest, moving inland, will send powerful winds from the high plateaus of inner Washington, Oregon, and Idaho across the Cascades and the Sierra to the coastal valleys. Since air warms as it descends (at about 5.5 degrees Fahrenheit per thousand feet), the temperature of the air masses may rise 30 degrees or more. Hot, dry winds roar down Sierra canyons and out across the Sacramento Valley, blow up bitter dust storms, whistle through the passes of the Coast Range, and flood the Bay Region with warm air. These "Diablo" winds (often approaching the bay from the general direction of Mount Diablo) are the equivalent of southern California's "Santa Ana" winds.

Fire Weather

The ocean fogs disappear in the face of the hot blasts from the interior. Sunbathers spread out on the lawns of the parks and flock to the ocean shore. For a few days, San Francisco has some of its rare beach weather. But this is also fire weather, and the lookouts on Tamalpais and other mountaintops anxiously watch the horizon for signs that the dried-out vegetation has ignited to start the inevitable brush fires.

With the coming of these hot, dry continental winds from the northeast, the humidity drops drastically and everything tends to dry out—clothes on the line as well as fruit on the trees and the skin on your face. Wild grasses already dried by the summer sun turn to tinder, ready to blaze at the first spark of a discarded match or cigarette. Even wooden houses, especially those with shingle roofs or walls, dry out and crackle in the heat, and in the cities the fire sirens wail persistently.

Sometimes a dry continental air mass moves down into the Bay Area more slowly, causing only a light breeze. San Francisco and other coastal places enjoy warm, clear weather without the desiccating effects of the northeast wind.

But it was a roaring northeaster that started a grass fire in the Berkeley Hills in September 1923 and fanned it down the dry slopes to the residential area, consuming hundreds of homes. Most of north Berkeley went up in flames, and the business district was saved only when the northeaster died out at the end of the second day and the cool sea breeze returned, turning back the holocaust. It was another blaze caused by a northeaster that consumed a great part of Mill Valley in July 1928, before being similarly turned back by the returning sea breeze.

In 1991, hot, dry winds from the interior fanned a blaze in the hills behind the Claremont Hotel on the Berkeley-Oakland border and destroyed more than 2,000 homes before the northeaster blew itself out after three days and the wind from the sea returned. The venerable resort hotel and thousands of homes to the south and west were spared by the change of aerial regime. These "fire winds" usually persist for two or three days (although fire weather has occasionally lasted for weeks), and San Franciscans with chapped lips and sunburned noses welcome back the cool, refreshing vapors from the ocean. On street corners, in downtown elevators, and in neighborhood grocery stores, the conversation is the same: "It's good to have the fog back again."

The Weakening of the Westerlies

When the jet stream stalls and the ridge or trough overhead stays put for long periods of time, the heat or fog may continue for weeks. When a trough settles in over the ocean, weakening the Pacific High and consequently diminishing the onshore winds and the normal upwelling, ocean-surface temperatures are warmer than usual—a matter of interest to fishermen as well as swimmers. Many kinds of fish that usually inhabit the cold upwelled waters off the northern California coast, including salmon and albacore, migrate northward and are not found in the usual numbers offshore. In their place, fishermen may haul in some warm-water fish—barracuda, bonito, yellowtail, and even an occasional giant marlin.

Some experts believe that the increased appearance of sharks off California beaches is owing partly to this warming of the waters in years when the upwelling does not take place. Conversely, it is possible that extraordinarily strong winds, caused partly or entirely by persistent ridges strengthening the Pacific High and increasing the upwelling, were responsible in the years from 1942 to 1956 for the disappearance of the great schools of sardines that had previously swarmed offshore and were the basis of a large fishing industry. (If so, those ridges responsible for the decline of the sardine catch also indirectly caused striking economic and social changes, such as the conversion of Cannery Row in Monterey into a major tourist attraction, including one of the world's most ingenious aquariums.)

The stalling of the jet stream affects not only swimmers and fishermen but the region's farmers. Long periods of warm continental air, resulting from a trough over the Pacific, will dry out the fruit on the trees and the vegetables in the fields. Normally, the summer fogs along the coast make the valleys opening onto the ocean ideal for growing cool-weather crops such as artichokes and brussels sprouts. The coastside areas of Santa Cruz and Monterey counties grow most of the artichokes raised in the United States. But when the Pacific High and the prevailing westerlies fail to produce the normal fog, the sun beats down on the coastal valleys, and the artichokes tend to dry out, with serious results for growers.

Livestock ranchers and dairy owners are also affected. Normally, the fog drip along coastal areas helps grow grass for grazing cattle. Without the fog, the grass is scanty, and ranchers have to buy more hay to feed their animals. On the other hand, when the fog penetrates too far inland and persists too long, it may spoil grapes in the Napa and Sonoma valleys.

Usually, however, the delicate balance of forces is maintained, giving the Bay Region, at the gateway between the land and the sea, its incomparable combination of continental and maritime climates, dryness and dampness, sun and fog, heat and cold.

The Gaps

At any particular point in the Bay Region, the line of demarcation between the two kinds of summer climate varies with the time of day, the distance from the ocean, and the proximity of "streamlines."

Mill Valley, for example, on the southeast side of Mount Tamalpais, may be damp and drizzly at night and early in the morning, but enjoy bright sun in the middle of the day. Yet it is far less foggy there in summer than at Stinson Beach, on the ocean side of the mountain. It has, in other words, more of a continental climate and less of a maritime climate than Stinson Beach.

Go eastward across San Francisco Bay and over the Berkeley Hills to Walnut Creek and the weather becomes steadily warmer and drier—more continental. Continue east over Mount Diablo to Byron and the marine influence is almost entirely gone. The temperature at Stinson Beach may be in the 50s at a time when Byron broils in the 90s.

In general, the same is true at any point in the Bay Region; a place farther from the ocean will tend to have less of a damp marine climate and more of a dry continental climate. Yet this principle is sharply modified by the streamlines along which ocean air is channeled through the hills. Two points about the same distance from the ocean may have widely varying climates because of their proximity to passes in the coastal hills.

To take an extreme example, Redwood City on the Peninsula, south of San Francisco, is about the same distance from the ocean as Berkeley, but it is normally warm and sunny on summer days when Berkeley is fogged in. The reason, of course, is that Berkeley is opposite the lowest gap in the coastal hills, the Golden Gate, while the Santa Cruz Range separates Redwood City from the ocean. The damp ocean air, channeled into a streamline by the Golden Gate, flows directly across San Francisco Bay into Berkeley.

The Golden Gate is the largest and lowest of the gaps in the Coast Range and has the greatest influence on Bay Region weather, but the other gaps function as "little Golden Gates." They funnel ocean weather inland along streamlines, sometimes allowing land weather to move to the coast, particularly during warm northeasters (see fig. 5).

One of these gaps is not far from Redwood City, causing the summer weather to be cooler than might be expected from its location, although not nearly so cool as exposed Berkeley. Just northwest of Redwood City, the San Andreas Fault slices through the hills from the ocean and has created a low-lying area—the Crystal Springs Gap—through which

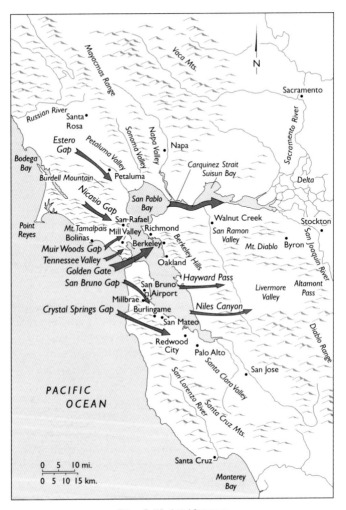

Figure 5. Wind and fog gaps

the breeze may penetrate in the afternoon and the fog may flow in the evening. The Crystal Springs Reservoirs, part of San Francisco's water supply, occupy part of this gap. The breeze through the Crystal Springs Gap not only cools Redwood City but may extend its refreshing influence as far south as San Jose.

Farther north is a much lower and broader pass between Montara Mountain and San Bruno Mountain, known as the San Bruno Gap, the historic and present route of El Camino Real. The San Bruno Gap is second only to the Golden Gate in its influence on Bay Region climate. The sea breeze and fog pour through this pass to cool off San Francisco Airport and the communities to the south, including Millbrae, Burlingame, and San Mateo.

Immediately north of the Golden Gate, there are three gaps: a narrow one at Tennessee Valley, a higher one above Muir Woods (both of which affect the climate of Mill Valley), and the considerably more important Estero Gap, which funnels winds and fogs from the Bodega Bay area into the Petaluma Valley. The cooling influence of the Estero Gap can be clearly felt if you drive north from San Francisco on a summer afternoon along Highway 101. When you round 1,600-foot Burdell Mountain and enter the Petaluma Valley, the temperature suddenly drops several degrees, and you feel the influence of the sea breeze. The Estero Gap (named for the estuaries at its seaward end) may extend its cooling effect as far as Santa Rosa, which may also be influenced by sea air moving through the Russian River canyon from the ocean.

Corresponding to these gaps in the westernmost range are three in the inner range just east of San Francisco Bay: Niles Canyon and Hayward Pass, which sometimes channel ocean breezes into the Livermore Valley; and the Carquinez Strait, the "inner Golden Gate," which channels the streamlines into the Delta and the Central Valley, lowering summer temperatures in Stockton and Sacramento.

There are many smaller gaps and passes affecting local areas. The same principle that affects the entire Bay Region holds true in San Francisco, for example; the summer weather is warmer with increased distance from the ocean, but that warmth is modified locally by streamlines (see fig. 6). One gap extends eastward from the beach along the line of Geary Boulevard, making that stretch of the Richmond District particularly breezy and foggy. Golden Gate Park lies in another gap; its streamline extends inland between Lone Mountain and Buena Vista Peak to the downtown area and blows the hats off pedestrians on Market Street. A long salient of fog moving slowly eastward through the park can often be seen from the surrounding hills. Marine air flowing over Twin Peaks can also result in downtown winds; often its fogs will evaporate as the air is warmed descending the eastern slope of the peaks.

By far the largest pass through the city, however, is the Alemany Gap immediately north of San Bruno Mountain. The wind and fog often flow from the Lake Merced area along the route of Alemany Boulevard, reaching San Francisco Bay near Hunters Point. One branch of the Alemany Gap extends through Visitacion Valley, channeling the streamlines toward the bay shore and around Bayview Hill to the old baseball stadium at Candlestick Point, where it collides with another streamline from the main Alemany Gap by way of Hunters Point, creating the world's windiest ball park. The convergence of streamlines sets up rapidly shifting eddies and pressure changes that have made a traveling baseball perform in unpredictable ways. Fly balls have been snatched by these vagrant gusts almost out of the mitt of many a frustrated outfielder. The aerial turbulence was a major reason the San Francisco Giants moved to their new stadium near the Bay Bridge.

Oddly, fog will sometimes flow massively through the Alemany Gap to the bay—and Candlestick Point—at times when all other areas around the city, even the Golden Gate, are sharp and clear. The explanation may be found in the varying directions of the wind off the ocean. If the wind comes from a direction to the north of northwest, the long tip of

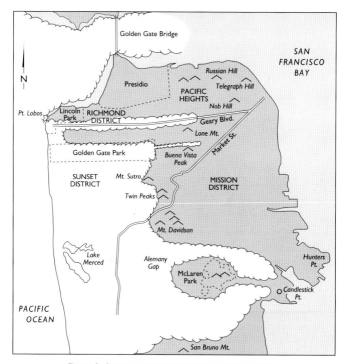

Figure 6. Some summer fog patterns in San Francisco

Point Reyes extending into the ocean thirty-five miles north-west of San Francisco may protect the Golden Gate, and the main force of the wind and fog will bypass that area and strike the San Francisco Peninsula southward near the city line, flooding in through the Alemany Gap.

Point Reyes may also to some extent protect the beaches at Bolinas and Stinson, making sunbathing there comfortable at times when San Francisco's Ocean Beach is whipped by stiff breezes. The two Marin beaches, particularly Stinson, may also be sheltered by Mount Tamalpais. Although Tamalpais is to the leeward rather than the windward, it rises so abruptly behind Stinson Beach, to an elevation of nearly 2,000 feet, that the beach is usually in a pocket of relatively dead air backed up against the mountain. The winds cannot whip directly across the beach as they do in San Francisco, where the backshore area is low and offers no protection.

FALL

The only dependable beach weather usually comes in early fall, which is San Francisco's real summer. Although elsewhere there is normally a lag of a few days to a few weeks between the summer solstice in June and the period of warmest weather, San Francisco's lag is unmatched in the United States; it amounts to about three months. September and October are the two warmest months of the year.

The reason is to be found in the workings of the same influences that created the summer fog canopy. After the summer solstice, the sun moves steadily south day by day. Its rays begin to slant through the layers of atmosphere, which weakens their impact. During the lengthening nights, the air cools off; during the shortening days, there is not enough time for it to warm up. As a result, by late September, maximum temperatures in the Central Valley may drop from above 100 degrees into the low 90s, and later into the 80s. In the High Sierra, late-season campers shiver in early-morning frosts.

The Departure of the High

Meanwhile, momentous changes are occurring over the ocean as well. During the summer, the Pacific does not warm up as much as the land, and it reaches its annual temperature maximum in early fall. Out over the rolling waters, the Pacific High has begun to move south with the sun and no longer sends strong winds racing toward the coast. The Central Valley now is not hot enough to suck drafts in through the Golden Gate and the other gaps in the Coast Range as it did in midsummer.

When the temperature difference between land and sea decreases, the push-pull effect that created the coastal winds and fogs no longer takes place. As the onshore winds die down, the upwelling that brought the cold bottom waters to the surface slows to a halt. The great coastal fog bank, which at its maximum reached perhaps 100 miles in width, now fades away to a few wisps along the shore. For a time, as the fog-forming process decelerates, the fogs may come through the Golden Gate low on the water, as they did in the early spring, forming similar fantastic shapes as they roll under the bridge, over islands and shores. Then they disappear entirely.

The sun, which during the summer was seen along the coastal areas only intermittently, now beams down in genial autumnal warmth, raising the San Francisco temperature into the high 70s and occasionally the 80s. Pale-faced residents again flock to the parks and beaches to lie in the sun and relax in a pleasant stupor. Surfboard riders in wet suits paddle out through the warming surf at Ocean Beach to take advantage of the glassy surface and the big swells that roll in from storms in the Southern Hemisphere, where the season is still winter or early spring.

The inversion layer of warm air that hung over the region during the summer—and was lifted at times well above the 1,000-foot level by the sea winds blowing in beneath it—now presses lower as the winds die down. Below it, the hazy surface layer of moisture-filled ocean air becomes steadily thinner over the land and dwindles to a few hundred feet. On Angel Island, the top of 780-foot Mount Caroline S. Livermore may rise out of the marine layer into the clear, warm air above, and even the tops of the Golden Gate Bridge towers may appear above the low layer of fog.

Smoke from chimneys around the shores of the bay rises straight up into the calm windless air, flattens out against the bottom of the inversion layer, thickens as it mixes with the moisture of the sea air, and combines with automobile emissions to become smog. Gentle afternoon breezes through the Golden Gate push the smog eastward until it is stopped by the Berkeley Hills and collects in the cities along the eastern shore and south to San Jose and the basin of the Santa Clara Valley. If the afternoon breezes do not materialize or are countered by sluggish movements of continental air from inland, San Francisco itself is wrapped in noxious gray vapors. On rare occasions, the smog over the bay has been so thick as to activate the fog horns.

Mirages

These low inversions of fall cause other unusual atmospheric effects. At times, the breeze through the Golden Gate may falter, then recommence, bringing thin layers of cool sea air in beneath the hazier, warmer air around the bay. The result is to lift the haze a short distance off the water. This process may occur a number of times in a few hours, until there are several layers of alternately hazy and clear air, all visible like layers of cake against the hills around the shore.

The bottom of an inversion layer sometimes has refracting qualities; the light bends and causes mirages. An apparent vertical elongation of objects on the horizon is a "superior mirage," an effect also known as "looming." Ships or clouds or the Farallon Islands may appear to be several times as tall as they actually are. Once, from the heights above the Cliff House, I saw the trees and rocks at Point Reyes, thirty miles away, so magnified and distorted that it seemed as though a skyscraper city was rising from the long peninsula. From Stinson Beach, a ship near the horizon may have an inverted image of itself directly above or below it. On one occasion, from a boat in Richardson Bay, off Sausalito, I saw an "inferior mirage" (in which the image appears below the object); the tall buildings of downtown Oakland appeared in the water north of the Bay Bridge.

Mexican Sky

The departure of the Pacific High with its onshore winds brings Bay Region residents the opportunity to observe some spectacular skyscapes quite unlike the fog displays of summer. They may originate thousands of miles away. Like the Atlantic hurricane belt that begins in the Caribbean and often moves northward, lashing the East Coast, there is a Pacific hurricane zone off Mexico. Although the Pacific hurricanes may occasionally strike the Mexican coast, they usually spin off to the northwest and fortunately expend their energy at sea. Their leftover humid air may bring high cloudiness and occasional thunderstorms to the Bay Region.

A less-frequent source of autumnal clouds and showers here may be the Gulf of Mexico, where masses of warm, moist air move north and west, bringing to the desert regions of New Mexico and Arizona intermittent thundershowers, locally called monsoon rains because of their similarity to the monsoons that deluge southern Asia. The results are sometimes flash floods, which are among the chief agents shaping the features of the desert landscape, particularly alluvial fans and deep gulches, which can be converted in a few minutes from dry creek beds to whitewater torrents.

The edges of these monsoon storms, like the more frequent outriders of Pacific hurricanes (or lesser oceanic storms) may reach the Bay Region and bring what might be called a Mexican sky—clouds of every description, producing evocative patterns of light and shade on the land and water. With an active imagination, you might at such times see herds of sheep moving cross the sky, stepping-stone designs, Spanish galleons, castles on the Rhine, celestial cities, all evolving into new shapes as they converge, separate, expand, and contract, as if they were produced by a master artist working in sky-hung mobiles. Thundershowers from either type of southern storm may bring lightning that can touch off wildfires in the mountains and set the Bay Region's hills reverberating with thunderous music of Wagnerian proportions.

As the sun continues seasonally southward, the nights grow longer and the days shorter. The atmosphere gradually cools, bringing bright fall colors to the bigleaf maples on the creeks of Tamalpais and a hundred other streams from Ben Lomond to the Russian River. The air is full of the spicy aromas of autumn. In the vineyards of Napa and Sonoma, the grape harvest is over and the leaves of the vines turn bright reds and yellows and purples.

But the weather is still unpredictable. The Big Game in late November may be held under a hot sun in a stadium that resembles an oven—or in brisk, snappy fall air appropriate for overcoats and the resounding thump of shoe leather on pigskin.

WINTER

Clouds

Equally as impressive as the Bay Region's flowing fogs of summer are the clouds of winter, reflected on the surface of the bay or changing shape and substance as they move over the surrounding hills.

It is unfortunate that clouds, like fog, have received a bad name in literature and legend. One dictionary definition of "cloudy" is "darkened by gloom or anxiety." A "cloudy day" and a "clouded countenance" are negative phenomena, to be avoided in favor of sunshine.

In reality, clouded skies can provide the highest form of nature's art, fantastic designs that if woven into fabric or duplicated on canvas would receive awed accolades. Either spread against the blue dome of the sky or chromatically lighted by sunrise or sunset, clouds can offer a moving spectacle, surpassing anything human art can create.

Beyond appreciating their esthetic value, the observer with some knowledge of cloudscapes can recognize signs of momentous happenings in the atmospheric regime, clues to the weather and the broad currents of air passing overhead. The beginning of that knowledge is proper nomenclature (see fig. 7).

Perhaps the most familiar types of clouds are **CUMULUS,** consisting of accumulations or clusters of visible moisture, flat on the bottom and billowy above. They are formed as a mass of unstable air, perhaps originating where the sun has heated an area of bare ground. The warmed air at the ground rises

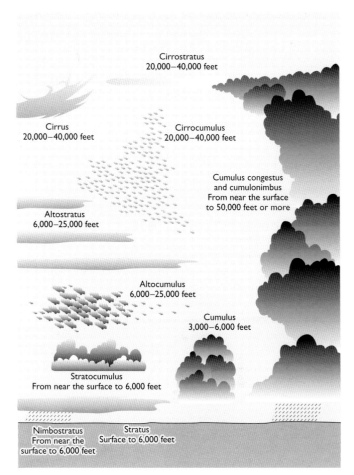

Figure 7. Cloud types (clouds at the same altitude appear here at different heights)

and cools, and its vapor condenses into visible form. Each cumulus cloud can be regarded as the top of a column of rising air. The flat bottom indicates the altitude at which the rising vapor has cooled to the point of condensation, and the clouds usually appear at 3,000–6,000 feet.

Low cumulus clouds take on individual shapes and drift across the sky with the breeze, in this region usually from the west. They are signs of clearing weather and are referred to as "fair-weather cumulus." These are the clouds that imagination can transform into herds of sheep or galleons sailing across the sky.

If the air is very unstable, cumulus clouds may boil upward into rising domes and battlements until they become thunderheads, or **CUMULUS CONGESTUS.** Forced upward by the approach of a cold front, they may tower as high as 50,000 feet, where their tops may flatten into an anvil shape, which spreads in the direction the storm is moving. If they produce rain, they become **CUMULONIMBUS,** the most spectacular of all rain clouds. They may form around the basin of the bay in late fall or early winter, rising above the surrounding mountains to reflect the sunlight in resplendent architectural structures that slowly change in shape. Sometimes they generate lightning and ear-splitting blasts of thunder.

Other types of clouds are classified by their shape and altitude. The lowest are the flat **STRATUS** clouds, the only kind that normally appear in this region in summer. Stratus forms in a layer of air that has been cooled uniformly (rather than in bunches or accumulations like a cumulus cloud). It may be produced when moist air contacts a cold surface. This is the process that occurs when an ocean wind passes over cold upwelled water, chilling the air and condensing its vapor, forming a fog bank. Or stratus can be created when clear air is forced to rise—for example, by encountering a mountainside—and cools until condensation takes place. This process, called adiabatic cooling, often occurs in summer when a clear sea breeze, blowing onshore, strikes a steep Coast Range slope and cools to the condensation point (dew point), developing a fog deck at elevations of several hundred to a thousand feet above sea level. It also occurs when clear marine air blows through the Golden Gate and across the bay, strikes the Berkeley Hills, rises, cools, and forms decks against the middle and upper slopes, leaving the lower air clear.

Plate 16. Each fair-weather cumulus cloud is the top of an invisible column of rising air that cools until its vapor condenses into visible form. The condensation usually occurs at altitudes of 3,000–6,000 feet.

Plate 17. Mid-level clouds (6,000–25,000 feet) form flat layers (altostratus) and clusters of ripples (altocumulus).

Plate 18. A setting sun transforms an altostratus formation into a chromatic skyscape.

Plate 19. Two altostratus formations seem about to collide as the cloud on the right sends out racing streamers of condensed moisture. Actually, however, winds are blowing the two formations in different directions at different altitudes.

Plate 20. A cumulus congestus forms east of San Francisco.

Plate 21. Cirrus clouds—high, thin veils composed of ice particles—may herald an approaching storm. They usually form at altitudes of 20,000–40,000 feet.

In the winter, stratus may form on the ground as radiation fog (see p. 70) or as clouds ("high fog") when air is forced to rise as it encounters a heavier air mass. Although usually flat, stratus may sometimes begin to billow out at the top as **STRATOCUMULUS.** If a stratus cloud produces precipitation, it is **NIMBOSTRATUS,** the ordinary rain cloud, a shapeless dark mass forming a uniform ceiling that may drop steady rainfall for hours. (In contrast, cumulonimbus rainfall is heavier but normally lasts less than an hour.)

In the middle altitudes, from 6,000 to 25,000 feet, are **ALTO-STRATUS** (high stratus) and **ALTOCUMULUS** (high cumulus). Both contribute to spectacular sunrises and sunsets. Higher yet, from 20,000 to 40,000 feet, are the **CIRRUS** clouds, transparent, gauzy veils casting no shadows, composed of ice crystals formed in upper regions where the air is below freezing. "Cirrus" means "curl," and often these clouds are blown by winds into feathery wisps or curls, sometimes called "mares' tails." If the clouds are flat or stratified, they are **CIRROSTRA-TUS.** If they accumulate in clusters or ripples, they are **CIR-ROCUMULUS**, sometimes called "mackerel sky" or "buttermilk sky." They, too, can be part of the color show overhead at the rising and setting sun.

Cirrus clouds indicate a high layer of air with a different origin and greater moisture than the air below. They may herald an approaching storm or may simply be the feathery edges of a storm passing far to the north.

Although the first rains to strike the region may come with the humid Mexican air in September or October, Indian summer often continues into November, which in coastal areas is usually warmer than breezy April. But meantime the jet stream with its storms has moved south with the sun and is in a position to affect California.

The storms are born out of the conjunction of warm and cold masses of air far out on the rolling surface of the ocean. A thousand miles off the China coast, a mass of cold air from the polar regions may move south over the Pacific, encountering a mass of warm southern air (see fig. 8a). The two

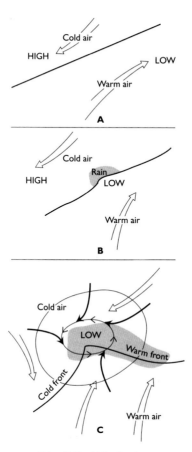

Figure 8. The birth of a storm

masses do not immediately merge, but may flow past each other in opposite directions. At some point, a bulge of light, warm air may rise over the edge of the heavy, cool air (see fig. 8b). As it rises, air moves in at the surface from all directions to replace the air that has risen.

The inrushing air, affected by the Coriolis force, cannot move straight into the low pressure area but curves to the right, causing the entire air mass to begin to move around in a counterclockwise direction, forming an eddy (see fig. 8c). This counterclockwise circling of air masses around a low-pressure storm center is called a "cyclone," although its winds are normally much slower than those of a tropical cyclone, or hurricane. At the center of the cyclone, the rising warm air cools and its moisture condenses into rain. A storm is born.

Storm Fronts

The entire revolving mass of air moves along the jet stream toward the Pacific coast of North America, continuing to draw fresh supplies of air into it.

If the cyclone at this stage of development were to pass over the Bay Region, the cool air already overhead would be succeeded by the warm air near the center of the cyclone, followed by another mass of cold air. Sometimes these air masses are visible as they pass overhead. The advance edge of each of these masses is known as a "front." Thus the cool air would be succeeded by a warm front, then a cold front. The entire process is sometimes called a "frontal system" (see fig. 9).

The warm air mass near the center of the cyclone, being lighter, tends to slide up over the edge of the cold air ahead of it. As it rises, it cools off, and its moisture condenses, creating a cloud. The form taken by the cloud depends on the height at which it condenses. The advance edge of the warm air mass, riding over the cool air, will attain heights of perhaps 20,000–30,000 feet. At this elevation, the temperature is so low that the moisture condenses into fine particles of ice, forming a cirrus cloud—the herald of a storm—a thin, hazy sheet of ice vapor barely visible high in the sky, perhaps forming a halo around the sun or moon.

As the storm moves on, the clouds lower and thicken to altostratus, then pass through regular stages to altocumulus, cumulus, stratocumulus, and finally the low, dark nimbo-

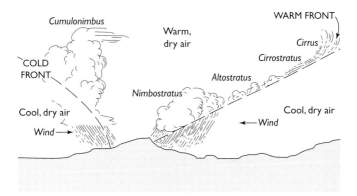

Figure 9. Cross section of a normal frontal system

stratus, or rain cloud. These later stages are usually not distinguishable by observers standing directly beneath them, but appear only as lowering, thickening clouds.

As the warm front at the center of the cyclone passes overhead, the air will be moist and full of mist, giving rise to what is called "hazy sunshine." Then comes the trailing cold front, marked first by nimbostratus and steady rain or towering cumulonimbus clouds and showers, then by clear cool weather as the clouds move on to the east.

Often, however, this "normal" sequence does not take place in the Bay Region. If the storm was born far out on the Pacific, as most of them are, it reaches the coast at a later stage of maturity. The trailing cold front tends to crowd in on the warm air mass at the center, scooping beneath it to join the cold air mass ahead. Thus the warm air mass is "occluded," or closed off (see fig. 10). It no longer reaches the ground but still drops its precipitation through the cooler air from high overhead. The normal sequence outlined above is abridged by

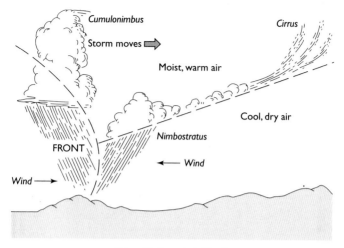

Figure 10. An occluded front

the elimination of the warm front, and the nimbostratus is succeeded directly by cumulonimbus.

Whether or not the normal cloud sequence appears at all depends on such variables as the amount of moisture in the storm, the depth of the storm, the stability of the air mass being lifted, and the circulation of high-level winds. A shallow storm, for example, will not reach altitudes high enough to form cirrus clouds and may be preceded only by the lower altocumulus. Or the appearance of cirrus may indicate the approach of a deep, moist storm bringing heavy rain. Cirrus may pass overhead, however, without heralding a storm; it may mark the edges of a storm passing far to the north or the remains of an old storm that has lost its energy at sea and is dying.

Often the coast will be hit by a series or family of storms, one behind the other, in such close succession that the weather does not clear between them. During the interval, the sky may be covered with various types of clouds, from low stratus to altocumulus.

Shifting Winds

In the Bay Region, cirrus is not always a good guide to coming storms, but there are other indications that are more reliable. The principal herald of a storm is a change in wind direction accompanied by low cloudiness. Flags on the skyscrapers of San Francisco, which are usually whipped stiffly eastward by the prevailing west wind, first droop listlessly on their masts, then begin to billow out toward the north. Rising smoke similarly drifts northward.

The reason for the wind shift is the counterclockwise circulation of air around the low-pressure center of an approaching storm. With the coming of a low from the ocean, the winds circling the low will come to the Bay Region from a southerly direction. The exact direction of the wind depends on whether the storm center is to the northwest or southwest. If the storm center is approaching the coast of northern California or Oregon, the winds will come at first from the southwest (see fig. 11). If the storm comes from the west or southwest, the first winds will come from the southeast.

As the storm moves on to the east, the direction of the wind changes to correspond with the stage of the storm. Watch flags on high poles or tall buildings; they act as weathervanes. If they fly out to the north, prepare for rain. When the rain comes, the south wind continues and is likely to bring the heaviest rainfall. As the storm progresses, the wind shifts, usually from south to west and finally to north. A west wind will be accompanied by lighter rainfall, and a north wind often brings intermittent showers as the low-pressure storm center moves on to the east.

The temperature also changes during these cycles of precipitation. As might be expected, the south wind, coming from warmer regions, is relatively warm; the north, as in tradition, is the source of cold. If there is hail or snow, it is likely to come with the frigid north wind.

I lived for some years in a house at the top of Telegraph Hill in San Francisco, where the course of a storm was easily

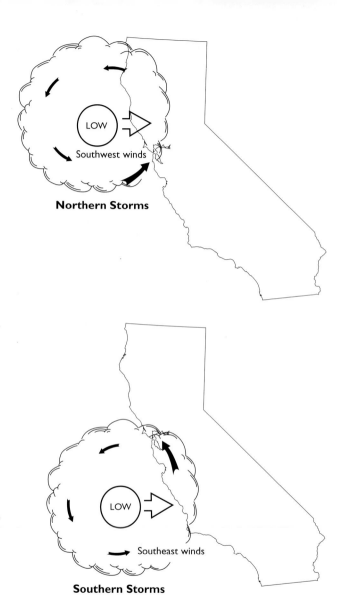

Northern Storms

LOW
Southwest winds

LOW
Southeast winds

Southern Storms

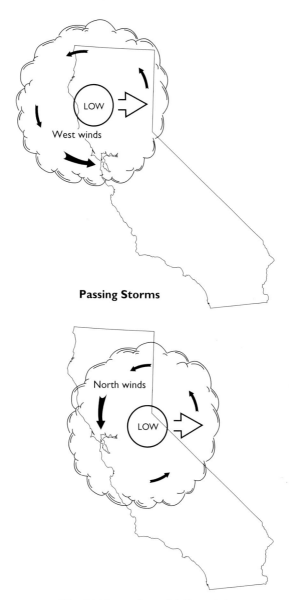

Passing Storms

Figure 11 *(above and opposite).* Storm winds

visible; raindrops on the south windows meant a storm in its early stages. But rain striking the west or north windows was an indication that the storm was passing away. The same test cannot always be applied to houses in lower areas, however, as hills or nearby buildings may cause wind eddies.

Similarly, it is sometimes possible to predict the weather by sniffing the wind. At the Telegraph Hill house, for example, a pungent aroma from the coffee roasters near the west end of the Bay Bridge was an indication of a movement of air from the south—and a possible storm. Sometimes the wind would come from a different southerly quarter, bringing the soapy smells of the copra docks at the southern end of the waterfront. (Unfortunately, in subsequent years, most of the coffee roasters and the copra trade left the city, making such olfactory predictions more difficult.) A salty smell meant an ocean breeze from the west through the Golden Gate, bringing fog or damp air off the ocean. The acrid scent of factory smoke indicated a movement of air from the industrial areas of the East Bay, primarily the refineries of Richmond and beyond. When the movement of air from the east and northeast was rapid, it would bring clear, dry continental winds from the high regions of the Great Basin—Nevada, Idaho, inner Oregon. When the movement was sluggish, it would sometimes be sufficient merely to counter the normal ocean breeze and carry the industrial smoke into the city, bringing smog.

Rain Patterns

The same conditions that cause a vast variation of summer climates within the Bay Region also create equivalent variations in winter climates. Just as certain communities get immense amounts of summer fog while nearby points are usually clear, so some places receive more rain than others.

San Francisco, for example, gets an average of 20 inches of rain annually, but Kentfield, nine miles north of the Golden Gate, is pelted with more than double that amount—48 inches.

Arid San Jose, with 14 inches, is only 20 miles from Ben Lomond, with 42 inches (see fig. 12).

The rainfall pattern is more complicated than the fog pattern. The summer fog comes from the ocean and dissipates to the east, but, as we have noted, the rain comes from several directions during the course of a single storm. Because the heaviest rain tends to fall during the early stages of a storm when the wind is from the south, the southern slopes of hills and mountains normally receive the greatest rainfall. Valleys open to the south winds will be wetter than those enclosed from that direction. For example, the San Lorenzo River flows through the Santa Cruz Mountains southward to the city of Santa Cruz and empties into Monterey Bay; because that river valley is open to the south, the rain-bearing south winds pour into it and drench communities such as Ben Lomond.

A similar phenomenon may take place in Marin County when the southerly winds collide with 2,600-foot Mount Tamalpais and are forced to rise, dropping their burden on the south side of the mountain, where Muir Woods National Monument records an average of 38 inches a year. The mountain rises so high above the bay that it snags passing storm clouds coming from any direction. The result is an average annual precipitation of 52 inches, the heaviest rainfall in the Bay Region.

San Jose is in the Santa Clara Valley, principally bounded by mountains on the south, east, and west, and is in the "rain shadow" of those ranges, protected from the rain-bearing southerly winds. If it were not for a narrow passage through the mountains to the south, San Jose would probably receive even less than its annual 14 inches.

In the later stages of a storm, as the wind shifts around to the west, the rain will come from that direction, pouring down on the western slopes of hills and mountains and leaving the eastern sides relatively dry. Half Moon Bay, on the seaward side of the mountainous San Francisco Peninsula, receives more rain than San Mateo or Redwood City, east of the same mountains.

The rainfall of a community is affected not only by its po-

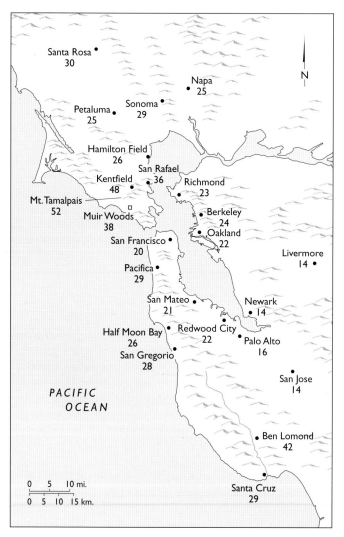

Figure 12. Annual rainfall averages (in inches)

sition in relation to nearby hills or mountains but also by its latitude. Because the jet stream is usually north of this region, storms are more likely to hit the coast north of San Francisco than south, striking Oregon and Washington, while only their southern fringes affect northern California. On occasion the Weather Service has been able to specify "rain north of the Golden Gate." The former Hamilton Air Force Base, on the northwest shore of the bay in Marin County, gets nearly one-fourth more rain than San Mateo, also on the shore of the bay, some thirty-four miles south.

Just as the Bay Region's rugged hill-and-valley contours cause wide variations in rainfall throughout the region, so similar contours on a smaller scale cause variations within a community itself, and the report of the community's weather station does not necessarily represent average conditions in the community.

Berkeley's rain gauge is on the University of California campus at the foot of the Berkeley Hills, at an elevation of several hundred feet, where the air has been forced to rise and deposit more moisture than falls on the lower parts of the city. In San Francisco, the precipitation reports that appear in the newspapers are based on observations made in one place—Duboce Park, near the geographical center of the city. But the city's hills and valleys may cause variations similar to those that exist on a larger scale in the Bay Region as a whole. More rain might be expected on the southern and western slopes of Mount Davidson and Twin Peaks—and perhaps even of Nob and Telegraph hills—than on the northern slopes; more in the oceanside Richmond and Sunset districts than in the hill-sheltered areas east of Twin Peaks. The Mission District, in the rain shadow of the hills partly surrounding it, might be expected to be drier than less sheltered districts such as the Western Addition or Pacific Heights.

Most of these possible variations—and those in similar microclimates throughout the Bay Region—have not been officially measured. They remain speculative and offer opportunities for amateur weather buffs to make their own discoveries.

The Pineapple Express

The types and amounts of rainfall here are strongly influenced by the direction from which storms approach along the jet stream. A storm center may move toward the coast from any part of the Pacific between the Gulf of Alaska in the north and the Hawaiian Islands in the south. Storms from the Alaska region begin with a southwest wind and tend to bring cold air. They occasionally bring snow to local mountaintops, from a few inches on Mount Diablo and Mount Tamalpais to a sugary sprinkling on Twin Peaks. The rare snowflakes that fall in San Francisco usually melt as they hit the ground, but on one frigid February day in 1887, residents were amazed and delighted when nearly four inches covered the rooftops.

The Alaskan storms bring cheer to the hearts of winter sports enthusiasts. They may leave snow as low as the 3,000-foot elevation on the Sierra Nevada, and at higher altitudes they usually deposit layers of dry powder snow, perfect for skiing. Hawaiian storms, on the other hand, usually begin with a southeast wind and are accompanied by warmer air. Because of the greater capacity of these warm air masses to hold water, they dump heavier rains on the lowlands and mountains as well. The Sierra snow line will thus be far higher—perhaps at 8,000–10,000 feet.

Warm Hawaiian storms, sometimes called the "Pineapple Express," are the despair of skiers, for their copious rains may turn ski runs to slush, and even at the highest elevations, they leave soft snow into which skis may sink to knee depth. It was a soft-snow, Hawaiian storm that buried and isolated Southern Pacific's deluxe passenger train "The City of San Francisco" near Donner Pass for three days in January 1952.

Hawaiian storms, because of both their greater water content and their ability to melt Sierra snows, bring the greatest flood danger. A series of such storms arrived just before Christmas in 1955, swelling the rivers to overflowing, inundating Yuba City, and deluging California with some of the greatest floods on record up to that time. For months afterward, the Sacramento and San Joaquin rivers spread a coating of muddy brown flood water across the surface of San Francisco Bay. Similar inundations from the Pineapple Express occurred in the 1960s, 1980s, and 1990s. Powerful storms may also result when the jet stream approaches from neither Alaska nor Hawaii but somewhere between, in the mid latitudes of the Pacific. In December 1995, for example, the storm track came from that direction and battered the Bay Region for thirty-six hours with torrential rains, accompanied by 100-mile-per-hour winds.

Tule Fog

In late fall or early winter—after the first rains have soaked the soil, the storms have departed, the nights are longer, and the air is clear, cold, and still—conditions are right for another change of mood in the atmosphere.

Just as the patterns of summer fog and winter rain at any particular point in the region are determined by its location in relation to the ocean and nearby mountains, so winter temperatures are strongly influenced by the same elements of location. The moderating influence of the Pacific is the greatest single conditioner of Bay Region temperatures. The ocean varies in temperature only a few degrees from winter to summer; consequently it tends to be cooler than the land in summer and warmer than the land in winter.

During frigid winter weather, land temperatures often drop far below those of the ocean. The warmest districts in winter are those along the coast. The coldest are inland valleys away from the ocean's direct influence—the farther away, the colder. In San Francisco itself, the oceanside districts, such as the Richmond and Sunset, are warmest; downtown and the Mission District, the coldest—a reversal of the summer pattern. Elsewhere, Half Moon Bay, for example, will be warmer than San Mateo, which in turn is warm compared to Walnut Creek. Coldest of all are the localities east of the Coast Range. Like the patterns of fog and rain, this winter temperature picture is modified by the passes in the hills and mountains, which tend to bring warmer coastal temperatures inland.

As the nights lengthen, the earth and the air quickly lose their heat after sundown, radiating it outward. The coldest air accumulates in the lower areas away from the warmer ocean. After the first winter rains, the cold air in these places absorbs moisture from the damp earth. In the frigid hours before dawn, the temperature sometimes drops so low that the moisture condenses, forming radiation fog.

Because the radiation fog develops most readily in low, damp places, such as the Delta, where tules and other marsh plants grow, it is commonly known as "tule fog." ("Tule," which rhymes with "truly," is an Aztec-Spanish name for bullrushes.)

The tule fog may form only a thin layer a few feet deep in the lowest areas. Sometimes the backs of grazing cattle can be seen eerily rising out of the mists. Pedestrians may walk down the road nearly blinded by thick vapors while the sun is shining at treetop level.

During a particularly cold, windless spell, the tule fog may accumulate night after night, until it is several hundred feet deep in some areas, particularly east of the Berkeley Hills. Because air usually tends to move from a cold to a warm area (from high pressure to low), the fog-bearing inland air begins to drift toward the warmer coastal regions. If the cold, windless weather continues for days or weeks, the thick white fogs pile up in deep drifts east of the Berkeley Hills, pour through the Carquinez Strait and other gaps in the hills, move across the bay to San Francisco and Marin, and roll outward through the Golden Gate, thus flowing in the opposite direction from the summer fogs, although at a much slower rate. At times, the tule fog may form on and around the bay itself and then move slowly oceanward.

A rough compensation takes place: inland areas that are usually fog-free in the summertime may in winter be deep in tule fog at a time when the coastal areas bask in clear sunshine.

Plate 22. Winter fog, commonly known as "tule fog" because it may originate in tule marshes, covers the valley floor under a broken ceiling of fair-weather cumulus.

Plate 23. A sea of tule fog, formed in the interior valleys and marshes during cold, damp winter nights, drains slowly down to the bay and moves westward (to the right) through the Golden Gate toward the ocean. Its motion is directly opposite that of the summer fog, which originates on the ocean and flows inland.

Plates 24 and 25. At sunrise, a dense tule fog blankets the bay from the Golden Gate to the Bay Bridge . . .

. . . silhouetting the towers of the city against a turbulent white sea.

In San Francisco, dark inland masses of moisture may bring gray gloom to downtown Market Street while out beyond the city's central spine of hills the oceanside Richmond and Sunset districts enjoy the sun.

As with the summer fogs and the winter rains, the gaps in the hills determine the local weather in detail. The Richmond District, nearest the Golden Gate, may be hit by a tule fog spilling over from the Gate, while the Sunset is clear. On other occasions, the fog may be thick enough to roll westward through the Alemany Gap, spreading over into the Sunset.

Sometimes the radiation fog reaches such depth and density that the San Francisco and Oakland airports are socked in; highway accidents reach a peak; ships probe gingerly through the murk. At such times, San Francisco Bay resounds with a symphonic orchestration of fog horns, bells, sirens, and ship whistles. Most shipping accidents, including the disastrous sinking of the liner *Rio de Janeiro* in the Golden Gate in 1901, with the loss of 130 lives, have occurred during thick tule fogs.

If the winter fog burns off under the heat of the morning sun, San Francisco becomes visible from the top downward. First appear the upper slopes of Twin Peaks, Mount Davidson, and other summits, rising like islands above a sea of white. Then the tops of the hotels on Nob Hill and the skyscrapers of the Financial District emerge from the mists; finally, the fog lingers only over the bay, where the funnels and superstructures of ships may appear long before the hulls of the vessels are visible.

Usually, tule fogs last for only a few days, then are driven away by the resumption of the normal ocean breeze, perhaps followed by another storm.

Just as in the summer, spring, and fall, the normal weather-making influences are upset when continental air masses move down to the Bay Region from the northeast, bringing dry winds and clear skies, so a similar phenomenon during the winter will abruptly replace normal weather conditions for a few days. The reason is global.

Rivers of Air

Over large parts of the earth, great rivers of air move in northerly and southerly directions as cold polar air masses flow toward the equator to replace the rising warm air of the tropics.

Over the continental United States during the winter, there are usually two or three rivers of air moving north and south. Often the largest of these is a current of cold Arctic air moving down from Canada. But the flow of cold air from the north must be balanced by a return flow of warm air from the south. Thus, one part of the country may enjoy sun bathing weather while another part freezes (see fig. 13a).

On occasion, southern air moving north may bring summerlike warmth to California in December, while east of the Rockies great currents of icy polar air flow down from Canada, bringing sharp, blue skies and sub-zero temperatures to many parts of the Great Plains. Winter temperatures in the Bay Region may rise into the 70s while the residents of South Dakota are having trouble chipping the ice off their windshields at 20 below zero.

Along the East Coast may come a third major current— warm, moist air moving up from the Gulf of Mexico. When this warm current encounters the cold air from the north, the result is usually spectacular—the birth of cyclonic storms bringing wind, rain, and snow across an arc of 2,000 miles from Albuquerque to Denver to New York to Boston.

At times, these north-south currents overflow their channels and spread perhaps a thousand miles in either direction. Icy Arctic air flowing down from Canada to the Mississippi Valley may swing west to the corridor between the Rockies and the Sierra (see fig. 13b). Then the Great Basin states—Utah, Idaho, Nevada—will have frigid weather while the Midwest basks in warmer air moving up from the south.

More rarely, the polar current of air will send icy tongues down west of the Sierra into the Central Valley and over the Coast Range, stirring up great clouds of dust in the Valley

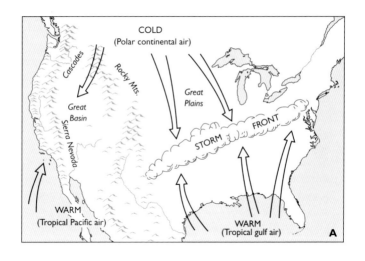

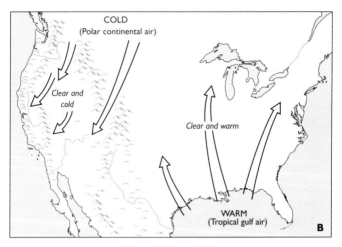

Figure 13. Rivers of air

farmlands, sending the temperatures down into the 30s, scuffing up whitecaps and frigid spray on San Francisco Bay, howling up Market Street, and sending shoppers and commuters scurrying for the warmth of stores and offices. Having been diverted west, the Arctic stream of air no longer chills the Great Plains, and newspaper readers in that region, enjoying unseasonal warmth, gloat over the cold spell in California. In December 1988, a frigid blast of polar air roaring across the Bay Region broke windows and toppled thousands of trees.

An invasion of polar air in the Bay Region may come two or three times during the winter. This brilliant, dry air brings the clearest weather of the year. On other occasions, the polar air mass does not blow into the region with great force, but seeps slowly over the ranges, bringing low temperatures, frost, and ice without clearing away the haze and fog. Usually, after a few days, the river of cold air returns to its normal channel and is replaced here by warmer marine air.

Violent Weather

The most violent atmospheric force to occur in the Bay Region—or anyplace on earth—is the tornado. Although tornadoes here are midgets compared to the giant twisters of the Midwest, they have nevertheless been powerful enough to cause serious damage.

Wherever they occur, they come in two varieties. Both are born in connection with thunderstorms moving along the jet stream. Some develop with giant (supercell) thunderstorms when air is sucked into the updraft and spins into a funnel-shaped whirlwind that may reach 200 to 300 miles an hour and cut a swath many miles long, pulling up everything in its path, including trees, roofs, cars, mobile homes, and other structures. (This is the kind that carried Dorothy to the Land of Oz.)

Others develop in conjunction with lesser thunderstorms, are shallower, less violent, and briefer, usually sustained for five minutes or less. Nevertheless, they can develop twisting winds up to 150 mph. The tornado, as distinct from other kinds of winds, wreaks its devastation in a narrow path where the tip of the funnel touches the ground. It moves across the land at a rate of 5 to 50 mph.

Most California tornadoes are of the lesser type, and they occur nearly every winter. But others can be larger and more disastrous. In December 1992, three supercell tornadoes roared through Santa Rosa, cutting broad swaths of destruction, and one day in December 1998, five to ten twisters ripped across the Bay Region.

As winter and spring advance into summer, the hot inner valleys are the scene of another kind of whirlwind far more common in California than the tornado—the dust devil. Under a blazing summer sun, a patch of barren ground gets much hotter than ground covered by foliage. Over the superheated spot a "bubble" of hot air forms, something like the bubble at the bottom of a pan of water on a stove.

The "bubble" finally gets large enough to burst loose and rise, leaving a partial vacuum below. Cooler air currents rushing in to fill the vacuum collide with such impact that they whirl into a twisting tornado shape, picking up dust, papers, leaves, sand, and sometimes small buildings. Fortunately, however, the dust devil lacks the tornado's devastating power.

The extreme heat that causes the dust devil does not normally last for many days at a time. In conjunction with the ridge-and-trough phenomenon passing overhead, the Valley heat draws ocean wind and fog through the Golden Gate and the other gaps in the low coastal hills, setting off the region's natural air-conditioning system. The ocean breezes gradually diminish the Valley heat, while the fog hangs thick over San Francisco Bay, setting off the great chorus of fog horns. After a few days of breeze, the Valley becomes cool again; the ocean air is no longer sucked in from the coast; and San Francisco Bay once again scintillates in the summer sun.

The earth continues its circuit around its blazing star; the rivers of air flow across the face of the planet in their summer courses; and here, where the Pacific has breached the mountains along the edge of the North American continent, the aerial forces of the land and the sea perennially collide in a visible war of the elements, ranging over mountains and valleys, through towns and cities, and across the face of the moving waters.

ACTIVITIES

The uniquely varied weather of the San Francisco Bay Region provides unparalleled opportunities for students and amateur observers to make valuable contributions to knowledge of the local weather and climate. The U.S. Weather Service is concerned with large-scale weather and cannot maintain stations to observe each of the region's microclimates—of which there are dozens within San Francisco alone. Systematic observations by amateurs in these areas would add significantly to an unexplored field of knowledge.

Particularly valuable studies can be made of summer fog cycles, correlating the fog conditions at various points in the region with temperatures at coastal and Central Valley points and with large-scale weather patterns available on the World Wide Web.

Another possible area of study is the rainfall pattern in various microclimates. How much more rain is there on one side of Twin Peaks, for example, than on the other sides?

Valuable aids in weather observation and prediction (obtainable at most scientific supply houses listed in the yellow pages of telephone directories) are a thermometer, barometer, rain gauge, and anemometer.

The barometer measures changes in air pressure. The most

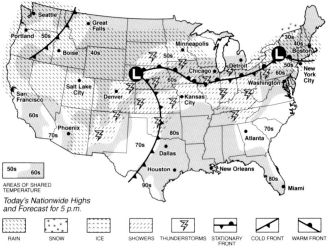

Figure 14. Daily weather map for April 6, 2001 (from the *San Francisco Chronicle,* courtesy of Weather Central)

common type is a mercury barometer—a tube about a yard long, sealed over at the top, with the bottom in a pan of mercury. If a vacuum is maintained in the tube, the outside air pushing down on the mercury in the pan will force the mercury in the tube up to a height of about 30 inches at sea level.

As the air pressure changes, the mercury will fluctuate up and down a few tenths of an inch. Cold, heavy air, for example, will press down with greater force on the mercury in the pan than warm, light air, thus sending the mercury in the tube up. The approach of cold weather is therefore characterized by a rising barometer. As a cyclone-type storm approaches with its lighter air, the barometer will fall.

The aneroid barometer registers changes in air pressure on a vacuum box. As the box top is pushed down by high pressure, the result is indicated on a dial.

Rain gauges are of many types, but the most accurate are those with a funnel collecting rainfall into a cylinder below.

The depth of the water in the cylinder will be greater than the amount of rainfall, depending on the proportion of the size of the funnel to the size of the cylinder.

The anemometer measures wind speed by a system of cups that rotate horizontally, indicating the speed on a dial.

Many newspapers print adaptations of Weather Service maps and forecasts, such as the one in fig. 14. Some television meteorologists display satellite observations of current weather conditions on regular news shows. The Weather Service and many other organizations publish up-to-the-minute weather maps on Web sites every hour. Some Weather Service stations can be visited by appointment. (For obscure bureaucratic reasons, the San Francisco Bay office is located in Monterey.) At several hundred small weather stations throughout California, volunteer observers collect daily records from barometers, thermometers, rain gauges, and anemometers.

SOME DEFINITIONS

HUMIDITY is the amount of invisible water vapor in the air. When air contains all the water vapor it can hold, it is saturated. Beyond that point, the vapor condenses and becomes visible as fog or clouds. The most common way to measure water vapor is in terms of relative humidity, which is expressed as the percentage of saturation. If air holds half the moisture it is capable of holding, the relative humidity is 50 percent. If air is saturated, the relative humidity is 100 percent. Because warm air is capable of holding more moisture than cold air, as the air cools, its relative humidity increases.

Owing to frequent ocean winds and fogs, the average relative humidity of coastal areas is high, except when the dry northeasterly winds bringing "fire weather" (see pp. 38–39) send the humidity down as low as 20 percent in San Francisco.

DEW POINT is the temperature at which condensation occurs when air cools and forms dew, clouds, or fog. When the temperature is low enough to freeze the vapor as it condenses, **FROST** is formed. In freezing weather, the dew point is the temperature at which frost will form; this is a vital matter to farmers, whose crops may be damaged. Frost is rare in the coastal areas because of the moderating influence of the ocean; it is more common in areas away from the ocean.

RAIN falls when water particles in the saturated air of a cloud increase in size and number until they cluster together to form drops. If the rain cloud is in freezing air at high altitudes, the drops will form as ice particles or snowflakes that will melt into raindrops when they reach the warmer air below. But if the air near the ground is very cold, they do not melt, and the result is a ground-level snowfall. Snow is rare in the San Francisco Bay Region—except floating in cirrus clouds high above. It seldom reaches lower ground, but in very cold spells, it may whiten such peaks as Tamalpais, Diablo, and Hamilton.

SLEET is formed when frozen drops in a cloud melt as they fall through warmer air below, then turn to ice again in a lower layer of freezing air near the ground. **HAIL** results when the ice particles in a thunderhead are carried high by the thunderstorm's updraft, adhere to each other, and form ice pellets that fall when they grow too heavy for the updraft. Sleet pellets are very small, usually less than three-tenths of an inch in diameter. A hailstone can be the size of a baseball, but the largest ones in this region may be more comparable in size to grains of rice. Both sleet and hail are unusual in the Bay Region, where the presence of the ocean keeps temperatures relatively warm.

LIGHTNING occurs when the turbulent air currents in a thundercloud cause a collection of positive electrical charges in one part of the cloud and negative charges in another. The attraction of the charges becomes so great that they rush together in a massive charge that shoots from cloud to cloud—or cloud to ground. **THUNDER** is the sound caused by the sudden expansion of the air heated by the lightning.

Because sound travels about a mile in five seconds, you can calculate your distance from the lightning stroke by counting seconds after seeing the flash. Start counting at the flash; if you hear the thunder five seconds later, the lightning is about a mile away; ten seconds, two miles, etc. If it is close, stay away from trees, flagpoles or other lightning attractors. Get into a closed car or a building, away from electrical or phone lines, wires, and metal plumbing. Avoid caves and gulches. If you are in the open, crouch as low as possible, with feet, hands, and knees on the ground.

IS THE CLIMATE CHANGING?

Yes, the climate is always changing. Look at the tree rings in the stump of an ancient redwood or bristlecone pine and you will see evidence that over the centuries there have been times of prolonged droughts and other long periods of very heavy rainfall. On the bottom of Lake Tahoe are the stumps of large trees that grew there when the lake level was much lower, indicating an extremely dry climate that persisted perhaps a century or more.

THE GREENHOUSE

In the last years of the twentieth century the question of climate change took on increased urgency. Do human activities cause climate changes that are much greater and much faster than those that have occurred naturally over previous millennia? Human actions contributing to the "greenhouse effect" and consequent global warming have become common topics of popular discussion and scientific investigation.

The greenhouse analogy is not totally accurate, but it is helpful as a rough metaphor. In an agricultural greenhouse, the sun's rays, coming through the glass, warm the soil, and the soil's heat is radiated upward. Instead of escaping into space, the heat is trapped by the glass and accumulates inside, raising the temperature. (Actually, most of the heating occurs because the greenhouse's interior is sheltered from cooling breezes—an inconvenient fact that we can disregard for the sake of the analogy.)

What happens on a global scale is partly similar. Sunlight warms the earth, and the warmth is radiated upward. In the atmosphere above, it encounters an accumulation of gasses that reflect much of the heat back to the ground, as a normal greenhouse roof would do. If it were not for these greenhouse gasses preventing the earth's heat from escaping into space, the planet would freeze over. Humans have reason to be thankful for the greenhouse effect, maintaining temperatures suitable for life.

That planetary greenhouse "roof" consists mostly of water vapor and carbon dioxide but also includes such gasses as methane and nitrous oxides. Human actions, mainly the increased burning of fossil fuels, such as coal and oil, since the beginning of the Industrial Revolution in the nineteenth century, have increased the amount of carbon dioxide in the global greenhouse by about one-third. Methane and nitrous oxides from agricultural fertilizers and other human sources have also increased.

The increase in the greenhouse gasses is roughly comparable to adding double or triple panes to the glass roof of an actual greenhouse, where the temperature rises as the thicker roof keeps more heat from escaping into space. Globally, the result, at least theoretically, is a rise in average temperatures. "Theoretically" because some scientists maintain that although average global temperatures have risen, the increase may be a result, not of human activity, but of natural cycles of climatic variation that have always occurred.

However, the Intergovernmental Panel on Climate Change (an international team of 2,500 scientists studying the problem for the United Nations) has concluded that human activities can be expected to warm up the earth. The scientists predict that if human production of greenhouse gasses continues, average global temperatures will increase by two to six degrees Fahrenheit over the next century. That amount may not seem like much, but small changes in average temperatures over time can have drastic effects. It was a mere nine-degree temperature drop that led to the ice ages of the past. An average global increase of two to six degrees (greater in some regions) would have catastrophic results in terms of a disruption of agriculture, depleting the world's food supply; greater floods and droughts; eradication of many species of plants and animals; and a rise in sea level (owing mainly to the melting of Antarctic ice) that could flood all the coastlines of the Earth and drive millions of people from their homes.

Critics of the global warming theory respond that the amount and speed of any sea-level rise have been exaggerated, and that in any case there would be plenty of time to build protective dikes. However, such measures could be prohibitively expensive, particularly in low-lying developing countries, and of doubtful feasibility. Locally, rising seas would require protection for low areas around San Francisco Bay. The Sacramento–San Joaquin Delta, which is part of the bay estuary, would be threatened; its numerous islands, with their rich agricultural soils, are already below sea level, and floods sometimes breach the levees around them. The Delta is the source of 65 percent of California's water supply; even a moderate sea-level rise would bring salt water into the channels and jeopardize the aqueduct intakes.

Ironically, global warming could conceivably create local cooling. Higher summer temperatures in the Central Valley might tend to draw more marine air in from the ocean, increasing the wind and fog along the coast. This effect would be muted if the Pacific itself grew unpredictably warmer. Otherwise, if global temperatures continue to rise, residents of coastal areas, including San Francisco and other shores of the bay, might expect to be fogbound for increasingly longer periods each summer.

THE OZONE LAYER

Another aspect of climate change made its appearance with the discovery of holes in the ozone layer, which is part of the stratosphere, concentrated at an altitude of about fifteen miles. Ozone is a colorless gas that exists not only in the stratosphere but at the earth's surface as a pollutant produced by sunlight interacting with auto exhausts. The ozone around us combines with other gasses and moisture to form obnoxious smog, but the ozone in the stratosphere is a benevolent shield that blocks most of the sun's dangerous ultraviolet radiation.

In the first few billion years of the earth's history, before the ozone layer existed, the ultraviolet radiation from the sun was so strong that life on land could not exist. But life in the oceans was sheltered, and about three billion years ago, the early marine life began to produce oxygen, which was released into the atmosphere. Ultraviolet light acting on the oxygen molecules ultimately created the ozone layer, conveniently blocking most of the deadly radiation and permitting life to evolve on land.

The discovery of "holes" in the ozone layer (more accurately a significant thinning of the ozone gasses) over Antarctica, and later over northerly latitudes, including North America, raised serious concern about the effects on life. Possible damage from a continued depletion of the stratospheric ozone includes a weakening of the human immune system, an increased number of skin cancers, and eye cataracts. Drastic disruption of natural and agricultural ecosystems would result if plants and trees were unable to endure the increased ultraviolet radiation.

Scientists reached the conclusion that the ozone-layer depletion is caused primarily by chemicals—chlorofluorocarbons, or CFCs—used in refrigeration, air-conditioning, and industrial processes and released into the atmosphere when the devices containing the CFCs are dismantled. Another source is the use of CFCs in aerosols, cans of spray under pressure. It is difficult to imagine that the release of chemicals from home refrigerators or from cans of hair spray could rise to the sky and weaken the shield protecting all life from lethal radiation. But enough scientists were convinced of that relationship to prod governments into action, leading to the Montreal Protocol of 1987, in which fifty-six nations agreed to phase out CFC production. By the turn of the century, the world was still far short of that goal. How the nations will comply with the treaty's requirements remains to be seen.

The CFCs already produced will remain in the stratosphere for decades. Even if the Montreal requirements were to be met on schedule, the ozone layer would not be restored to its previous level until the mid twenty-first century. The increased ultraviolet light makes it prudent for anyone to avoid prolonged exposure to the sun. Even in foggy San Francisco and other coastal regions, ultraviolet rays can penetrate an overcast sky. Sunbathers beware!

EL NIÑO

The effects of global climate changes on the Bay Region can only be understood in the context of wind and weather patterns over the entire Pacific. Although Robert Louis Stevenson's reference to "trade winds" in San Francisco was inaccurate, recent scientific discoveries indicate that the trade winds do indirectly have a significant connection with Bay Region climate, even though they are thousands of miles away over the tropical Pacific.

In a broad band on both sides of the equator, where the sun's rays are strongest, the heated air rises. To replace the rising air, cooler air rushes over the ocean toward the equator from the north and south. Owing to the Coriolis force (see p. 9), the wind from the north curves to the right, or clockwise, turning west. In the southern hemisphere, the Coriolis force causes the moving air to turn to the left—counterclockwise—also blowing to the west (see fig. 15a).

The combined result of these two moving air masses on both sides of the equator creates persistent winds from east

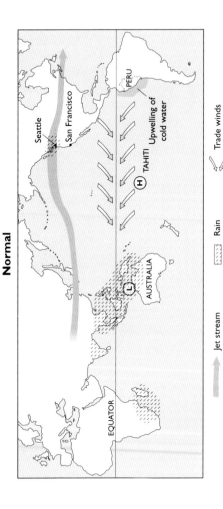

Normal

Figure 15a. In the "normal" air circulation over the Pacific, trade winds, rushing toward the equator to replace rising sun-heated air, curve westward as a result of the Coriolis force, causing an upwelling of cold water off Peru. The winds are pushed farther west along the equator by a high-pressure zone centered over the Tahiti region and attracted by a low over northern Australia. As they travel over the ocean, they pick up more moisture, which is dropped as monsoon rains in southern Asia.

Jet stream   Rain   Trade winds

El Niño

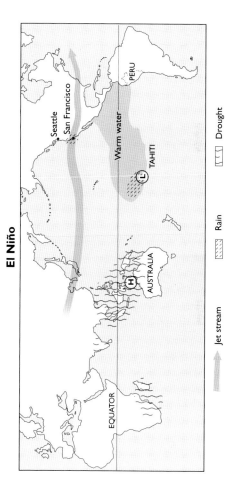

Figure 15b. During an El Niño episode, the situation is reversed, with a low over Tahiti and a high over Australia. The trade winds die; the upwelling stops; the ocean surface warms up in the eastern Pacific. The jet stream over the North Pacific, which normally brings storms to Oregon, Washington, and British Columbia, moves south, picking up warmth from the warm-water bulge below, and deluges California.

to west in the tropics, an effect felt as far north as Hawaii, where the climate is dominated by the winds from the northeast. These are the trade winds that filled the sails of merchant ships crossing the Pacific from North America for trade with Asia. (They returned across the northern Pacific with the prevailing westerlies.)

These easterly winds push the surface waters of the ocean along with them, resulting in an east-west current across the central Pacific. As the air and water flow westward, they are further heated by the sun; the moist air rises, cools and drops its water content in monsoon rains on the far side of the Pacific, from Indonesia to India.

Meanwhile, along the coast of Peru, just south of the equator, the sun-warmed surface currents have been pushed offshore in summer by the trade winds (much as the surface currents off California are moved offshore by the northwest winds coming down the coast). The westward-moving surface water must be replaced by cold water from below—the upwelling that brings nutrients feeding great stocks of commercially valuable fish.

That's the "normal" picture. But there are years when the trade winds diminish or die. The upwelling off Peru slackens or halts, and the water begins to warm. Without the trade winds, warmer water from the open ocean may spread toward the continent, reversing the usual direction of the equatorial surface currents (see fig. 15b).

This heating effect occurs to some degree every four to seven years, and because it usually begins around Christmas, South American fishermen refer to it as "El Niño"—the Christ child. But it is scarcely a blessing or a gift for them; they dread it, because it diminishes their catch. In the absence of the rich nutrients that are brought up from the bottom by upwelling, the cold-water species depart, to be replaced by less commercially valuable warm-water species.

The dying of the trade winds has another effect: without the great quantities of moisture picked up from the surface and borne along on the westward-moving winds, the heavy rains that they normally cause in the western Pacific, particularly in Southeast Asia, no longer arrive. The result can be drought from Australia to India, where the lack of monsoon rains causes great losses in agriculture.

But nature always compensates in one way or another. Drought in these regions means more rain elsewhere. Warmer waters, and consequently warmer air in the eastern Pacific, induce more evaporation, more rain clouds, and more precipitation along the Pacific coasts of the Americas, including California.

As the trade winds weaken and the ocean off South America warms up, the accompanying low-pressure zone there acts as a remote "magnet," attracting the jet stream with its storms southward from its usual route across Canada and the Northwest. The storm track may cross California, bringing the type of very wet weather that is normally found in Seattle and Vancouver. Those cities are then likely to experience some extraordinarily dry weather. In the Bay Region, the alternating pattern of ridges and troughs moving with the jet stream from west to east will bring intervals of dry weather between storms.

In the big El Niño episodes of 1982–83 and 1997–98, the Bay Region received more than twice the normal rainfall. Giant waves from storms born in the warm Pacific waters battered sea cliffs north and south of the Golden Gate. In Pacifica, clifftop dwellers saw their homes undermined and swept into the ocean. Elsewhere, houses were destroyed by mudslides, bridges were washed out, and highways were blocked.

Although El Niño events occur every four to seven years, they vary greatly in timing and strength. A mild El Niño will scarcely have any important effect, but a strong one can bring disaster. The outlook for El Niño episodes in the twenty-first century is uncertain. If global warming continues, bringing warmer air and water, the result may be that El Niño events will increase in frequency and intensity.

LA NIÑA

The opposite of El Niño is the less-well-known La Niña. Although you might suspect that the feminine version would be warmer and less violent than the masculine, the reverse is likely to be true, at least in California. The tempestuous La Niña occurs when the trade winds are stronger than usual, pushing more sun-warmed surface waters westward, causing the upwelling of more cold water off South America, and the intensification of the circulating oceanic currents of the North Pacific. The wintertime effect of La Niña in the Bay Region is likely to be colder, windier weather and maybe abnormal rainfall in either direction, too much or too little (and sometimes neither), depending on the erratic location of the jet stream. If La Niña persists into the summer, stronger upwelling off California brings more summer fog to the Bay Region.

SCIENCE AND THE UNKNOWN

I have been describing the El Niño-La Niña phenomenon as a result of the decrease or increase of the trade winds in the equatorial regions, but the question remains: Why do the trade winds vary so widely? Are their variations truly the ultimate cause of El Niño, or are they the effect of some other phenomenon?

Probably both. Every cause is an effect of some other cause or combination of causes. Meteorologists describe the weakening of the trade winds as a result of an extraordinary high-pressure zone in the western Pacific, around Australia, for example, and low pressure over the central Pacific, as in the vicinity of Tahiti. This is the reverse of the "normal" situation, in which there is a low over Australia and a high over Tahiti. These reversals of high and low in the Pacific are known as the El Niño Southern Oscillation—or ENSO (see fig. 15b).

At any rate, the Tahiti low accompanies El Niño episodes and tends to attract the jet stream farther south over North America, as we have seen, bringing to California the rains that normally drench Oregon, Washington, and British Columbia. The cause of this periodic oscillation of the high- and low-pressure zones in the Pacific is obscure. The best that meteorologists can do is to shrug and refer to "reasons that are not yet completely understood." Or they may refer to chaos theory, which holds that some phenomena are so complicated as to be unpredictable.

Such is the nature of science. Trace any cause-and-effect pattern far enough and you ultimately venture into realms where human knowledge has barely begun to penetrate. There are innumerable puzzles yet to be deciphered, questions yet to be answered, mysteries yet to be solved. What science does know as certainty might be compared to a drop of water in a near-infinite ocean. All the progress of science up to the twenty-first century is only the first stage of the heroic human venture, begun thousands of years ago, to understand the meaning of the world around us.

FURTHER READING

Bernard, Harold W., Jr. *The Greenhouse Effect*. 1980. New York: Harper & Row, 1981.

Blumenstock, David I. *The Ocean of Air*. New Brunswick, N.J.: Rutgers University Press, 1959.

Calder, Nigel. *The Weather Machine*. New York: Penguin Books, 1977.

Ehrlich, Paul R., and Anne H. Ehrlich. *Healing the Planet: Strategies for Resolving the Environmental Crisis*. Reading, Mass.: Addison-Wesley, 1991.

Graedel, Thomas E., and Paul J. Crutzen. *Atmosphere, Climate, and Change*. New York: Scientific American Library, 1997.

Gribben, John, ed. *The Breathing Planet*. New York: Basil Blackwell, 1986.

Huning, James R. *Hot, Dry, Wet, Cold, and Windy: A Weather Primer for the National Parks of the Sierra Nevada*. Yosemite Natural History Association / Sequoia Natural History Association, 1979.

Kals, W. S. *Your Health, Your Moods, and the Weather*. Garden City, N.Y.: Doubleday, 1982.

Kimble, George T. H. *Our American Weather*. Bloomington: Indiana University Press, 1955.

Knox, Joseph B., ed. *Global Climate Change and California: Potential Impacts and Responses*. Berkeley and Los Angeles: University of California Press, 1991.

Laskin, David. *Braving the Elements: The Stormy History of American Weather.* New York: Doubleday, 1996.

Lee, Albert. *Weather Wisdom.* Garden City, N.Y.: Doubleday, 1977.

Ludlum, David M. *The American Weather Book.* Boston: Houghton Mifflin, 1982.

Lynch, David K., ed. *Atmospheric Phenomena: Readings from Scientific American.* San Francisco: W. H. Freeman, 1980.

McAdie, Alexander. *The Fogs and Clouds of San Francisco.* San Francisco: A. M. Robertson, 1912.

Reifsnyder, William E. *Weathering the Wilderness: The Sierra Club Guide to Practical Meteorology.* San Francisco: Sierra Club Books, 1980.

Schaefer, Vincent J., and John A. Day. *A Field Guide to the Atmosphere.* Boston: Houghton Mifflin, 1981.

Schneider, Stephen H. *Global Warning: Are We Entering the Greenhouse Century?* San Francisco: Sierra Club Books, 1989.

Stewart, George. *Storm.* New York: Modern Library, 1947. A classic novel of a storm moving through northern California in the 1940s, with authentic weather details.

Suplee, Curt. "El Niño / La Niña." *National Geographic,* March 1999, pp. 72–95.

Thompson, Phillip D., and Robert O'Brien. *Weather.* New York: Time-Life Books, n.d.

Watson, Lyall. *Heaven's Breath: A Natural History of the Wind.* New York: Morrow, 1984.

Williams, Jack. *The Weather Book.* New York: Random House, 1982. Excellent graphics.

Young, Louise B. *Earth's Aura.* New York: Knopf, 1977.

INDEX

Page numbers in *italic* refer to figures; page numbers in **bold** refer to plates.

Illustrator:	Bill Nelson
Indexer:	Patricia Deminna
Compositor:	Integrated Composition Systems
Text:	9.5/12 Minion
Display:	Franklin Gothic Book and Demi
Printer and binder:	Everbest Printing Company

11/0 5   5   y/04
  11/10 (7) 9/08
4/15 (8) "/11